How to Publish a

Kindle Book

with Amazon.com:

Everything You Need
to Know Explained Simply

By Cynthia Reeser

HOW TO PUBLISH A KINDLE BOOK WITH AMAZON.COM: EVERYTHING
YOU NEED TO KNOW EXPLAINED SIMPLY

Copyright © 2011 Atlantic Publishing Group, Inc.
1405 SW 6th Avenue • Ocala, Florida 34471 • Phone 800-814-1132 • Fax 352-622-1875
Web site: www.atlantic-pub.com • E-mail: sales@atlantic-pub.com
SAN Number: 268-1250

Library of Congress Cataloging-in-Publication Data

Reeser, Cynthia, 1977-
  How to publish a Kindle book with Amazon.com : everything you need to know explained / by
Cynthia Reeser.
       p. cm.
  Includes bibliographical references and index.
  ISBN-13: 978-1-60138-404-1 (alk. paper)
  ISBN-10: 1-60138-404-1 (alk. paper)
  1. Electronic publishing. 2. Kindle (Wireless reading device) 3. Electronic books--Publishing.
4. Self-publishing. I. Title.
  Z286.E43R44 2010
  070.5'73--dc22
                                        2010005991

PROJECT MANAGER: Kim Fulscher • PEER REVIEWER: Marilee Griffin • ASSISTANT EDITOR: Holly Marie Gibbs
PRE-PRESS & PRODUCTION DESIGN: Samantha Martin • INTERIOR DESIGN: Harrison Kuo
FRONT & BACK COVER DESIGN: Jackie Miller • millerjackiej@gmail.com

Printed on Recycled Paper

We recently lost our beloved pet "Bear," who was not only our best and dearest friend but also the "Vice President of Sunshine" here at Atlantic Publishing. He did not receive a salary but worked tirelessly 24 hours a day to please his parents. Bear was a rescue dog that turned around and showered myself, my wife, Sherri, his grandparents Jean, Bob, and Nancy, and every person and animal he met (maybe not rabbits) with friendship and love. He made a lot of people smile every day.

We wanted you to know that a portion of the profits of this book will be donated to The Humane Society of the United States. *–Douglas & Sherri Brown*

The human-animal bond is as old as human history. We cherish our animal companions for their unconditional affection and acceptance. We feel a thrill when we glimpse wild creatures in their natural habitat or in our own backyard.

Unfortunately, the human-animal bond has at times been weakened. Humans have exploited some animal species to the point of extinction.

The Humane Society of the United States makes a difference in the lives of animals here at home and worldwide. The HSUS is dedicated to creating a world where our relationship with animals is guided by compassion. We seek a truly humane society in which animals are respected for their intrinsic value, and where the human-animal bond is strong.

Want to help animals? We have plenty of suggestions. Adopt a pet from a local shelter, join The Humane Society and be a part of our work to help companion animals and wildlife. You will be funding our educational, legislative, investigative and outreach projects in the U.S. and across the globe.

Or perhaps you'd like to make a memorial donation in honor of a pet, friend or relative? You can through our Kindred Spirits program. And if you'd like to contribute in a more structured way, our Planned Giving Office has suggestions about estate planning, annuities, and even gifts of stock that avoid capital gains taxes.

Maybe you have land that you would like to preserve as a lasting habitat for wildlife. Our Wildlife Land Trust can help you. Perhaps the land you want to share is a backyard— that's enough. Our Urban Wildlife Sanctuary Program will show you how to create a habitat for your wild neighbors.

So you see, it's easy to help animals. And The HSUS is here to help.

2100 L Street NW • Washington, DC 20037 • 202-452-1100
www.hsus.org

Author Acknowledgment

I would like to extend my appreciation to the editorial staff of Atlantic Publishing Group, Inc. for their professionalism and keen eyes, and to the case study participants who deigned to discuss their opinions and industry knowledge with me for this book. This book is dedicated to the advancement of technology and its contribution to the progression of the sometimes-fusty, old field of traditional publishing.

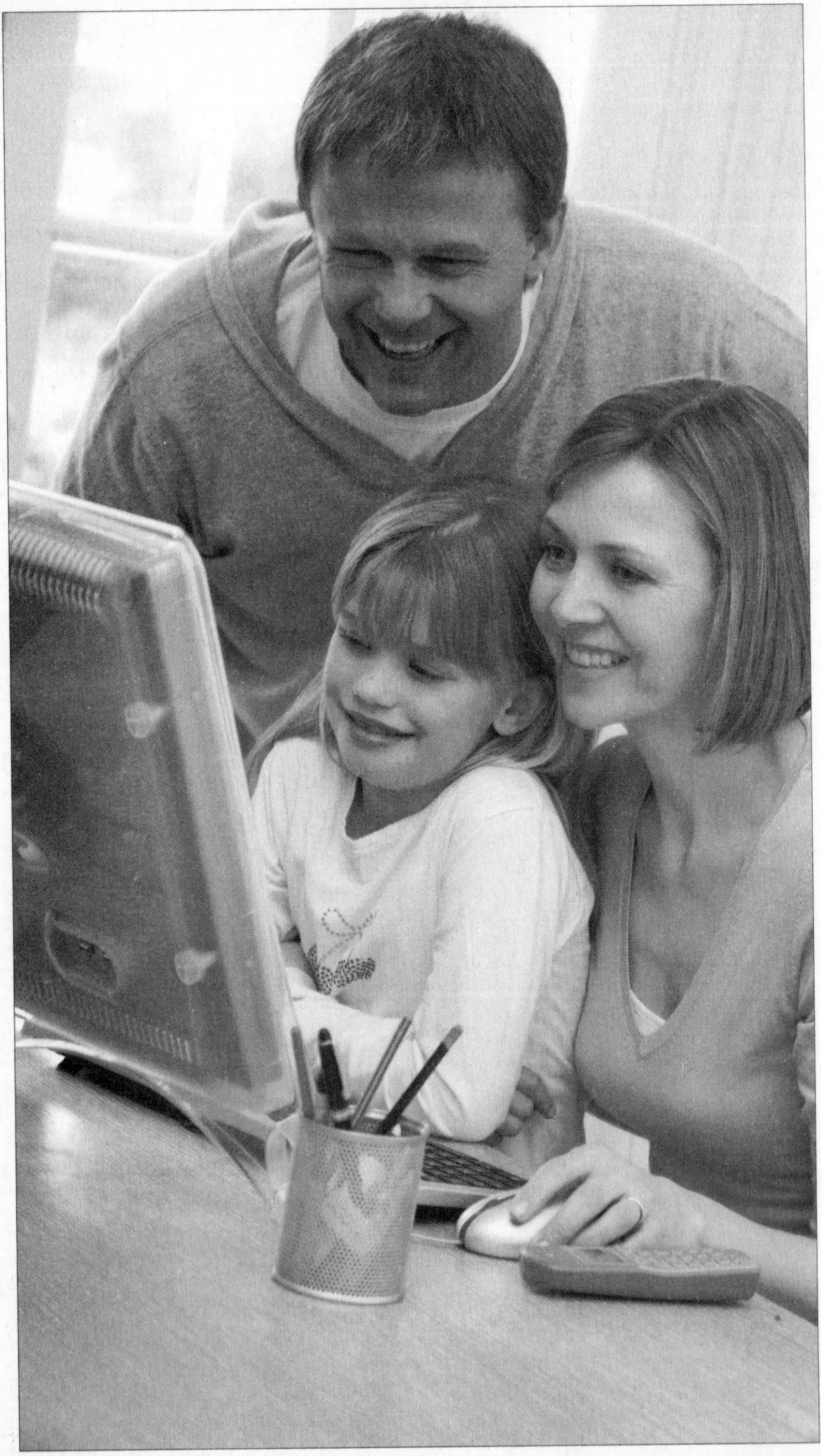

TABLE OF CONTENTS

CHAPTER 2: GETTING STARTED: AMAZON AND THE DIGITAL TEXT PLATFORM

CHAPTER 5: PUBLISHING FOR MOBIPOCKET AND MOBILE DEVICES 127

CHAPTER 6: KINDLE PUBLISHING FOR BLOGS, NEWSPAPERS, & MAGAZINES 153

CHAPTER 7: MARKETING AND PROMOTION 175

CHAPTER 8: MAKING THE MOST OF THE INTERNET AND SOCIAL MEDIA 191

CHAPTER 9: REVIEWS AND MANAGING PUBLISHED WORK   211

CHAPTER 10: GENERATING IDEAS FOR NEW CONTENT   223

CHAPTER 11: UP TO DATE: STAYING ABREAST OF CHANGES IN THE INDUSTRY 241

CONCLUSION 255

Publishing, the concept of a book, and the way we read are at the heart of a changing literary landscape — and there is no reverse. It is not evolution — suggesting gradual, peaceful, and progressive change over a sustained period — it is revolution. Successful, bestselling authors are self-publishing eBooks; retaining control and electronic rights over their work. More traditionally published authors will follow the lead of Stephen R. Covey, Piers Anthony, Anne Rice, J.A. Konrath, and others.

What becomes essential in this do-it-yourself publishing environment where independent authors have freedom and control is 'how-to' information. In *How to Publish a Kindle Book with Amazon.com: Everything you Need to Know Explained*, Cynthia R. Reeser delivers a timely 'full-cycle' explanation of every aspect of the Kindle publishing process. However, *How to Publish a Kindle Book With Amazon.com* goes way beyond this; it is a complete, indispensable guide for all authors — whether independent or traditional, publishing in print or eBook — on all aspects of publishing including publicity, sales, editing, social media, managing expectations, staying ahead of the game, self-employment, taxes, and accounting.

Foreword

When the first Kindle launched in 2007, few appreciated the significance of the moment that would revolutionize the publishing industry, and make publishing accessible to the masses. Just a few years later, a new-model Kindle witnesses triple digit growth in e-book sales. Since most of us failed to appreciate the potential of the Kindle and eBooks in 2007, perhaps it is not so wild now to speculate that the eBook will do more than change the nature of books and reading, but that it will save rainforests and species under threat. Adding to the mix are the 'now' generation: children growing up in a Jetson world where gadgets and convenience reign supreme. They are not attached to the idea of a printed book or the pastime of bookstore browsing. For them, going to a bookstore to buy a book is much like our grandparents visiting their local store to buy fabric, sugar by the bag, and hatpins. For this tech-savvy generation, reading may become trendy again with the ease in which books can be downloaded to a Kindle, its portability, and its various features and accessories.

Utilizing relevant and recent case studies by experts in their fields, *How to Publish a Kindle Book with Amazon.com* delivers practical, experiential advice and explanations on key concepts, such as the technology behind the Kindle and the elements at play in this changing landscape. *How to Publish a Kindle Book with Amazon.com* is a thought-provoking reality check for those who remain skeptical despite mounting evidence that eBooks, dominated by its leader, the Kindle, are the future, and authors really can do it themselves. It is a revolution, and I — we at the Association of Independent Authors — are pleased and excited to be a part of it.

Regards,
Leigh K. Cunningham
Executive Director
Association of Independent Authors
www.independent-authors.org

You know something approaching a revolution is happening in the publishing world when the sales of electronic books surpass the sales of print books. Yes, it actually happened. For the first time, according to an article from Dec. 28, 2009 on MSN Money Central, purchases of eBooks for the Kindle® exceeded purchases of books in print format from Amazon.com®. In a Dec. 26, 2009 news release now available on Amazon.com, CEO and founder Jeff Bezos thanked customers for "making Kindle the most-gifted item ever in our history." While Amazon was not forthcoming with specific sales figures, it is evident from the rising shares that the economy according to Amazon is on the up-and-up. Amazon shares rose 1.7 percent from the yearlong period from Jan. 7, 2009 to Jan. 7, 2010. There is no doubt about it: this is the information age, and reading is hot.

And speaking of reading being hot, in a by-now-oft-quoted-interview with Charlie Rose, when Jeff Bezos was asked *Why the name Kindle?* he responded, "To start a fire." And start a fire the Kindle has certainly done. Prior to the Kindle, there were a few different versions of eBook readers, with varying degrees of usefulness, complexity, and popularity. There was, for example, the 1998

Introduction

Rocket eBook from NuvoMedia, and there were eReaders being produced by Sony, such as the 2003 Librié. But none of these devices had quite the vision or scope of the Kindle. Many people have asked, *Why the Kindle?* Certainly, why the Kindle over other eReaders? The answer to that question is multi-faceted:

The Kindle can take credit for being a device that has brought eBooks, which certainly existed before even the first eReader, into the hands of more users than any like device before it. Perhaps largely due to its association with Amazon, the largest online retailer of books, the Kindle has been responsible for the sale of more copies of eBooks than any other device or retailer. Add the Kindle's convenience and ease of use to the equation, and you have an instant success story. Not only can the current version of the Kindle store 1,500 books, but it also uses an Internet technology that does not require additional fees, and that has recently been enabled for global wireless accessibility. The larger version, the Kindle DX, stores 3,500 books. The technology of E-Ink is a huge bonus, too: think of E-Ink as polarized black and white particles that arrange themselves to form the words on your screen, which does not require backlighting, or an internal light, to illuminate pixels — all of which equates to greater readability, even in direct sunlight.

Even aside from the innovative technology of the device itself, Amazon has opened up the Digital Text Platform (DTP) to enable anyone to publish and sell his or her work for the Kindle. That means that books from publishers like Harper-Collins™ are available in Amazon's Kindle Store right alongside books from independent publishers and small presses. But arguably the most exciting aspect is the way the Kindle (and Amazon) is changing publishing as an industry. There is heated debate at the time of this writing about how publishers — now that they are taking eBooks seriously on a fiscal level — are going to handle electronic versions of an author's work, especially where rights, royalties, and the overall contract is concerned. And perhaps best of all, authors with true talent who perhaps, for one reason or another, could not attract the attention of a large publisher, now have a platform to both have their work recognized on its own merit — apart from any publishers, agents, or contracts — and to earn royalties outside of a publishing contract. Of course, even authors with little talent can publish for the Kindle; the process is democratic, not monopolistic. And for those reasons, the Kindle is going a long way toward changing the negative connotations of self-publishing and the stodgy ways of traditional publishing.

So much in the publishing industry is changing, and the infamous Google^{SM} Book settlement has been yet another hot topic at the forefront of the eBook revolution (if such it may be called). In 2005, The Authors Guild and The Association of American Publishers brought a copyright infringement lawsuit against Google, known as *The Authors Guild, Inc., et al. v. Google Inc.*, Case No. 05 CV 8136 (S.D.N.Y.). Also involved in the class-action lawsuit were individual authors who brought objections against Google's Book Search procedures of scanning copyrighted materials to make them available to the public through Google Books (**http://books.google.com**).

An initial settlement was proposed, and an amended settlement was later filed on Nov. 13, 2009. The primary changes resulting from the amended settlement included the creation of an independent fiduciary (the Book Rights Registry^{SM}), adjustments to the representation of foreign countries, the management of unclaimed funds, the support of out-of-print works, payment for unauthorized scanning of copyrighted books and documents, and greater control for authors and publishers, among others. A primary focus of the settlement is to put the control of works back in the hands of copyright holders. Authors (and authors' estates) who hold the rights to their work control the degree of the content Google can make available to the public through a Book Rights Registry account management page.

As a result of the action, members participating in the lawsuit stand to receive not only compensation for books they own copyrights to that were scanned in to Google Book Search, but also revenue from future scanning. The new nonprofit organization, the Book Rights Registry, was founded as a result of the settlement and serves as an independent collector and disburser of funds and revenues collected from third-party content users. The Book Rights Registry maintains a database that keeps track of rights holders for published works and positions itself to resolve disputes that may arise among rights holders. Libraries were the primary sources of scanned works for Google Books; now they may be designated in multiple categories:

1. **Fully participating libraries** sign an agreement with the Book Rights Registry that frees them from copyright infringement liability, provided they follow certain guidelines. A library under this type of agreement scans its in-copyright books for Google, and Google provides the library with an electronic copy of each scanned work. Interestingly, what is known as a library digital copy (an LDC) can be used to create

a new print version of the document in the event it is damaged, lost, stolen, deteriorating, or destroyed.

2. **Cooperating libraries** are provided with a copyright infringement liability waiver from Google so they are protected in case they destroy Google's copyrighted LDCs. These types of libraries provide copyrighted works to Google for purposes of scanning, but generally do not retain Google's LDCs.

3. **Public domain libraries** provide Google with works in the public domain only, and Google in turn provides them with a copyright infringement liability waiver, similar to that described above.

4. **Other libraries** have opted out of the Google settlement, but provide Google with scannable material.

For more information on the Google Book Settlement, see **http://wo.ala.org/gbs**.

Besides the fact that Google Books makes print materials available in electronic format to users who might not otherwise have access to the work — as much of the available material comes from public and university library holdings — another important feature is the availability of out-of-print works. With the issue of copyright (mostly) addressed, the general public is gaining access to what essentially amounts to a worldwide library collection. It is a researcher's dream, and an important step to the accessibility of information. Now, print works are being integrated onto the Internet en masse; until fairly recently in its history, the Internet was primarily a collection of digital resources. The Internet is growing as a vast repository of information.

That repository is becoming increasingly democratic in its process. With access to increasing amounts of information that vary in source and quality, people are presented with a steadily growing range of choices. Because technically, anyone with Internet access can now publish for an eReader or mobile device, the public is forced to develop more sophisticated filters and ways of discerning quality. But the Internet itself is a repository of information whose reliability and accuracy of sources and information spans a wide spectrum of quality and usefulness. When it comes to eBooks, the principal is the same: not every eBook is just as useful, accurate, or helpful as the next.

While writing this book, I decided it would make sense to publish an eBook of my own. After all, what better way to research the methods of publishing for the Kindle than to do it myself? As the editor-in-chief of a quarterly online literary journal, *Prick of the Spindle*, I simultaneously published our first print edition, the "Fiction Open Competition No. 1", which is the result of our first fiction contest, and our first foray into print.

Contemporary reading life in most developed nations is informed on a regular basis by both print and online mediums, with print in its hard copy form gradually taking a backseat to its electronic counterpart. It is beginning to seem like everyone is a writer, with more working professionals with widely diverse areas of expertise writing blogs all the time. In fact, according to **www.blogcount.com**, there were an estimated 2.4 million to 2.9 million blogs considered active in June 2003. The Internet & American Life Project at the Pew Research Center™ estimated that the blog population in the U.S. to be at 12 million adults in July 2006. Since 2002, there have been some 133 million blogs indexed (catalogued) by Technorati℠ at **www.technorati.com**. So it is no wonder that the Kindle Store now offers blog subscriptions. At the time of this writing, there are 5,000 blogs available on the Kindle. A visit to the Kindle Blog and News Feed main page (access this by navigating to **www.amazon.com**, then going to the Kindle Store and scrolling down to Blogs & News Feeds, then clicking on the link, "See all Kindle Blogs & News Feeds") reveals blogs indexed in categories such as Arts & Entertainment, Business & Investing, and Internet & Technology, to name a few. The magazines available in the Kindle Store at the time of this writing number 46.

Naturally, as the editor of a journal, I was curious. My issues are quarterly, but could I not find some way to publish them for the Kindle? Also, with so few magazines on board the ePublishing wagon, what if I got a jump-start on the competition? Moving from our primary format as an online publication to one available on the Kindle seemed a natural move. On Nov. 30, 2009, a mere week before Amazon started their beta program for periodicals, I e-mailed **digital-publications@amazon.com** with my query. How could I go about publishing my quarterly issues for the Kindle? Exactly one month later, I received an e-mail. It was an invitation to the private beta program that began Dec. 7, 2009. The benefits, the e-mail indicated, would consist of an "intuitive interface" for feed configuration, and assistance from a team of integration managers. It came with a link and a brief set of instructions. How exciting!

I had not held out much hope that I would hear from anyone, but there it was: a private invitation, along with a phone number and private contact address. I promptly signed up for a Kindle Publishing account, and went through the process of formatting and submitting an issue of my literary journal to the Kindle Publishing for Newspapers and Magazines platform. Pending approval by Amazon, regular issues of the journal will be available in the Kindle Store, available through Amazon.com. *In Chapter 6, I describe my experience and provide detailed instructions on the formatting and submission process.*

So far, the process has been easy and rewarding, barring some potential barriers of technical knowledge (of XML) that reared their heads during the process of formatting for the Kindle Publishing for Newspaper and Magazines platform. Using the Digital Text Platform, formatting and posting the title are the easy parts; after the writing and development of the content, the biggest challenges come with marketing and promotion.

For those of you who have already published work traditionally, why bother publishing anything for the Kindle? Because the Kindle is yet another avenue of publication — one whose market is quickly expanding, as the holiday season of 2009 proved when it was dubbed the most gifted item ever. Or, because the Kindle encourages the sale of electronic books over print books; extra revenue and exposure certainly never hurt an author.

The Kindle has indeed started a fire, and it is one that has grown larger than anyone, perhaps even Jeff Bezos, ever expected it to. Or, maybe some kind of publishing revolution is what Mr. Bezos and his team of developers had in mind all along.

Author's Note: As you read this book, please keep in mind that information online changes very rapidly. It merits mentioning that, as I was writing this book, the information changed so rapidly that I found myself going back frequently to update details on Web page navigational directives, changes to the devices themselves, and other similar items. By the time this is in your hands, some or many of the details may have changed. Therefore, I encourage you to make use of the Internet as a tool to discover the information you may need to feel in the gaps, and to use this book as a resource in guiding you along the process of publishing your material electronically.

Liberation from Traditional Publishing Standards

The traditional method of getting published through a mass market or trade publisher is a lengthy and often arduous process. Anyone who has ever poured their heart and soul into a manuscript — taking it carefully through the process of editing, only to have it rejected by agents and publishers — understands how difficult it can be to get printed by one of the big names in publishing. While it is valid to continue developing and shaping your work so it is the best it can be, new methods of publication have emerged in the last decade that are redefining the face of publishing. Large publishers like Random House™ and Houghton Mifflin are seeing the appeal of ePublishing, most directly in their pocketbooks.

EPublishing is a method utilized increasingly by small and large presses alike. The emergence of publishers whose entire collection of works consists solely of electronic publications is another relatively recent publishing phenomenon. While authors must usually go through the same process of carefully editing and otherwise pre-

CHAPTER 1:
Power to the people:
The Revolutionary Amazon Kindle

paring and submitting their work, the decision process with ePublishers is often faster, with the chances of publication leaning more in favor of the author. Why? The answer is simple: time and money. Methods of electronic publishing save on both.

Amazon.com's handheld reading device, the Kindle, is no exception. With the added value of convenience and marketing exposure through the online giant Amazon.com, it is no wonder the Kindle has received such favorable attention. The liberation the Kindle provides is perhaps most clearly evidenced by its tendency to help level the playing field: it includes selections from large publishers, small presses, and independent authors alike.

A Brief History of Electronic Publishing and Digital Books

EPublishing takes a variety of forms; including CD-ROM, print-on-demand (POD), and eBooks, which can be downloaded on a computer or a handheld reading device, such as the Kindle. According to economist and Fordham University professor Albert N. Greco, the collective line-up of products and services that comprise a digital publishing platform and appropriately equipped information age office were positioned for revolution by the late 1990s. That lineup includes the Internet, eReaders and handheld reading devices, eBooks, cable, fax, wireless services, and the like.

Greco views the burgeoning of electronic print in terms of displacement — that is, the displacement of print material. From the mid 1990s through the late 1990s, the media was awash with the notion that electronic media (commonly referred to as "new media") and e-devices would soon phase out print media altogether. However, a decade into the 21st century, the predominant e-reality is indeed the more balanced notion of displacement, in the sense that eMedia represents the next major shift after the development of the book in print or the printing press. While Kindle sales continue to increase along with the device's popularity and the acceptance of the electronic book format, books in their print and audio versions have devoted audiences and appear to be in no danger of extinction.

By the late 1990s — at the end of a century that had seen the most technological advancement in the entire history of mankind — electronic publishing entered into existence as an industry. Writers and others in the publishing industry encouraged a widespread switch to electronic readers and ePublishing, touting the most likely reasons the shift was so highly anticipated: durability, portability, and readability. Some scholars cited interactivity and ease of cross-referencing as key advantages of eMedia.

Figures from a decade-long study (from 1996 to 2006) by Communication Industry Forecast reflect a direct correlation between the increase in consumer time spent using the Internet and electronic media, and time spent reading printed material. EMedia is undoubtedly upsetting print media from its former throne, but the change is proving to be gradual, and not instant, as some eMedia enthusiasts had formerly posited.

CASE STUDY: MUSINGS WITH SCREENWRITER/ KINDLE PUBLISHER JOHN AUGUST

CLASSIFIED CASE STUDIES™
directly from the experts

Screenwriter: *Big Fish, Charlie and the Chocolate Factory, Corpse Bride, Titan A.E., The Nines, Go, Charlie's Angels*

Has Published on Kindle: "The Variant," a short story

John August's screenwriting credits include Go, Big Fish, Titan A.E., Charlie's Angels *(2000), and* Charlie's Angels: Full Throttle *(2003),* Charlie and the Chocolate Factory, Tarzan, *and* Corpse Bride. *He also maintains a screenwriting-oriented Web site at **http://johnaugust.com**. Born and raised in Boulder, Colorado, John earned a degree in journalism from Drake University in Iowa and an MFA in film from the Peter Stark Producing Program at the University of Southern California. He lives in Los Angeles. John has a weekly screenwriting column on IMDb (Internet Movie Database), in the "Ask a Filmmaker" section of **indie.imdb.com**.*

Interview conducted via e-mail

Question: Could you talk about your background as a screenwriter, and how you came to write a short story? Was the story originally an idea for a screenplay?

Answer: The underlying story that became "The Variant" was originally conceived to be a screenplay. I never wrote it as one, but that was my initial intention. A friend had asked me to write a short story for an anthology he was planning, so I dusted off the idea and wrote it as a short story. Even as I was working on it, I was thinking about the possibility of also selling it through the Kindle platform.

Q. Could you discuss your experience in publishing for the Kindle?

A. I'm a relentless do-it-myselfer, so I read up on other people's experiences with getting books on the Kindle and just tried it. The formatting isn't that complicated if one's done HTML [*See*

Appendix B for a basic HTML guide], and the account setup is pretty straightforward for anyone who has experience managing Web sites and registrations.

Q. Are you a Kindle owner? If so, could you talk about reading eBooks versus reading print books?

A. I've had a Kindle since almost the beginning, and will almost always prefer to read something on it rather than printed books. It's more convenient, but it's also one less thing I have to bring into my house. Books have a way of stacking up. I don't feel the same psychological weight with a book on my Kindle.

Q. What is your process like as a writer; how do you develop your ideas?

A. In most cases, ideas just force their way into my head and insist that I pay attention. They're like fascinating-but-insistent new acquaintances, spilling all their secrets. After a few days, I get sick of most of them and send them packing. But others hang around for a long time, bubbling up to reveal something new. Eventually, many of those ideas get worked into stories. That's the craft of it — figuring out how to tailor good ideas into actual narratives with beginnings, middles, and ends.

Q. Do you have a favorite eBook?

A. My favorite book is usually the one I'm reading. I'm reading *Thinking in Systems* by Donella Meadows right now, which is great.

Meet Kindle: A Brief Introduction to Amazon's Kindle Electronic Reader

The first Kindle edition was launched on Nov. 19, 2007, but it was not the first eReading device on the market; a much earlier eBook predecessor was the Rocket eBook, released by NuvoMedia in 1998. After the original stock of the first

Kindle sold out in five and a half hours (some sources say it was six), scores of backorders were taken, and the stock was not replenished until April 2008. Two years later, Kindle launched their third edition — the DX.

Kindle

The first Kindle impressed users with its long battery life (the battery requires recharging approximately every other day when the device is left on), ability to store a great deal of content, and obvious convenience. The design, however, did not go over so well with some users. Initially retailing for $399, the price was soon lowered to $359. The Kindle was, unfortunately, only available to U.S. consumers because its Whispernet (Internet) technology was functional only in the United States. The Kindle stores about 200 books, but can hold more with the addition of a memory card.

Kindle 2

Amazon.com announced the Kindle 2's debut on Feb. 9, 2009, and made it available for purchase on the 23rd of that month. The cost was initially $359, but was lowered to $299 on July 8, 2009, and has since been lowered to $259. The second edition of the Kindle features a 6-inch diagonal screen, weighs in at a little more than 10 ounces, and features 2-GB internal storage and a slimmer profile than the first version. One battery charge holds power for the device for four days with wireless turned on, and for two weeks with wireless off. Battery life is approximately 25 percent longer than the prior version.

The EVDO modem (Evolution Data Only/Evolution Data Optimized) allows wireless Internet access without relying on WiFi. Whispersync™ allows users with more than one Kindle device or those using the Kindle app on their iPhone™ devices to automatically update all devices on the last viewed page in the book they are reading. Users are not obligated to any contracts and do not pay wireless connection fees. Wireless coverage is global, and an international version of the device debuted in October 2009.

The Kindle 2 holds more than 1,500 books and includes an automatic library backup. Image resolution is improved, and text size is adjustable. A five-way controller makes flipping through magazines and newspapers easy. The text-to-speech feature enables users to have the book read to them — a convenient option for anyone who is traveling, occupied, or visually impaired. Additional features include a Web browser, built-in dictionary, and the option to transfer MP3 files to the Kindle, which you can play as background music while reading. A USB connection is included for use with the Kindle power adapter, or to connect to a Mac or PC. An optional "book cover" (protective case for the device) is sold separately. The cost of covers ranges from $9.99 to $139.99.

Kindle DX

The $489 Kindle DX is larger than the Kindle 2, and boasts more storage capacity. The screen measures 9.7 inches diagonally, and the device has a 4-GB internal storage capacity and holds 3,500 books. While still ultra-lightweight, it gains a few more ounces than the Kindle 2, coming in at 18.9 ounces over the Kindle 2's 10.2-ounce frame. Like the Kindle 2, a single battery charge is good for 4 days with wireless on, and up to two weeks with wireless off, depending on accessibility of coverage. In areas where wireless coverage is low, more battery power is consumed. Also like the Kindle 2, wireless coverage is provided via Amazon Whispernet℠ (Whispernet is a name Amazon gives their Internet functionality), which originally utilized the Sprint 3G network, but now uses the AT&T℠ 3G network.

The same standard features that are included on the Kindle 2 are also present on the DX, such as the text-to-speech feature, MP3 capacity, built-in PDF reader, dictionary, and Web browser. The primary advantage the DX has over the Kindle 2 is screen size and storage capacity. The larger screen means no scrolling to see a page in its entirety, and the screen auto-rotates for vertical and landscape views for easy perusal of maps, graphs, and charts. The auto-rotation feature for the DX is said by users to be somewhat clunky. According to a June 12, 2009 review on the device by Harry McCracken in The Technologizer Review, this feature "sometimes worked correctly, albeit sluggishly; but it often didn't flip when I wanted it to, or flipped when I would have preferred that it didn't." The auto-rotation feature can be disabled.

What Kindle Provides for Readers

Kindle offers distinct advantages over traditional reading and publishing. These are some of the most important factors:

Convenience

A reader can condense an entire library in a handheld device weighing a few ounces. The Kindle is as thin as a magazine, and is easily portable. With the advantages of accessible-any-where, always-on wireless via Amazon.com's Whispernet and no contract or monthly access charges, the Kindle is the ultimate handheld eReading tool. The accessories included with the Kindle, such as the text-to-speech feature, basic Web browser, access to Wikipedia, Whispersync, built-in diction-ary, and MP3 and audio support, are included bonuses. The Kindle is ideal for anyone — busy professionals, students, avid readers, writers, researchers, and just about everyone else. With the number and genre of titles increasing constantly, even children may find some enjoyment in the Kindle.

Discount pricing

When most cutting-edge electronic devices debut, they are priced high, and gradually see some decrease in cost. The Kindle was no exception. The original debuted at $399, and later was lowered to $359; the Kindle 2 debuted at $359, and then was discounted to $299. Now the price has dropped even further, to $259, and Amazon.com sometimes has refurbished models available for purchase at a lower price. Currently, the refurbished models are few and far between, but as new mod-els are released in the years to come and the popularity of the device continues to increase, a greater number and variety of refurbished, discounted models should appear on Ama-zon.com. The recently released, larger DX model costs $489, but is sure to see a drop in price eventually. As with most any device, more people will make the purchase when the price is lowered, therefore increasing the popularity further.

Prices as of November 1, 2010		
Kindle Wi-Fi : $139.00	Kindle 3G: $189.00	Kindle DX: $379.00

Diversity of content

Titles in just about every imaginable genre on just about every imaginable topic are already available for the Kindle. Books are available in horror, mystery, best-selling fiction, memoirs and nonfiction, business, legal, self-help, how-to, children's, and many more genres. With the launch of the DX, the educational and newspaper markets began to take notice. The larger format of the DX makes it appealing for use as an educational device, and the space-saving (and potentially cost-saving) appeal to college students may be a no-brainer.

Already, Amazon.com is partnering with three educational publishers who make up about 60 percent of the market for textbooks — that is, Pearson Education, Inc., John Wiley & Sons, Inc., and Cengage Learning. Also on board with the partnership are Princeton University, Arizona State University, Case Western Reserve University, Reed College, and the University of Virginia, who began testing the program in Fall 2009. According to an article on CNET.com, Case Western President Barbara Snyder said students would be observed for changes in study habits and their learning processes.

Newspapers in print have long been experiencing dwindling readerships and, as a result, are commonly shifting focus to online content. Including newspapers in downloadable Kindle format just makes sense — it is a natural next step for an industry that many were beginning to think was a dying breed. Not surprisingly, *The New York Times*, *The Boston Globe*, and *The Washington Post* were three newspapers that decided to test the waters on the Kindle during the summer of 2009 by making their product available for the Kindle in areas without home delivery. According to a May 7, 2009 *Los Ange-*

les Times article, the Kindle DX, which had just debuted the day prior, established a partnership with New York Times Co. and Washington Post Co. to allow anyone living outside of the delivery areas for *The New York Times*, *The Washington Post*, and the *Boston Globe* to purchase the Kindle at a discounted rate with a newspaper subscription. According to a media analyst quoted in the article, the move was a last-ditch effort by newspapers to generate new readerships. Last ditch or not, the idea caught on. With 93 newspapers in the Kindle Store at the time of this writing and newspaper subscriptions available on competing devices, the newspaper industry has found a new (and expanding) delivery outlet.

The Kindle could turn out to be more than just a convenient tool — it has the potential to be at the forefront of a publishing revolution. When you stop to consider what the printing press did for the world (dissemination of information, literacy, and communication), consider what a bold step the Kindle constitutes. If entire libraries can be carried around in a purse or day planner, what impact might such a level of convenience mean? No one knows just yet.

Refunds

Refunds for eBooks may be offered for various reasons — sometimes an author uploads a book but does not preview the document, and it goes to the store for purchase, errors included. Some books are incomplete, due to an error during uploading. Sometimes, a technical glitch occurs, like garbled and unreadable text (this can happen when material is converted from one file format to another, such as when Mobipocket files are converted to Kindle files).

Amazon's refund policy is listed as a link at the bottom of the Amazon.com home page. Scroll all the way down to the bottom right hand corner, to the column that reads "Let Us Help You"; under that, click "Returns" to be taken to the Returns and Refunds page. To view the Kindle Returns Policies, scroll down to the "Product-Specific Return Policies" and click the link for Kindle. See "Returning Kindle Content" for specific information. This subheading informs users of what looks to be a fairly lenient returns policy for downloaded content.

Anything purchased from the Kindle Store is eligible for "return and refund" as long as the customer makes her request within seven days of the purchase. After the refund, the download will be removed from the customer's Media Library and she will no longer be able to read it. Customers wishing to request a refund go to the Contact Us link in Customer Service, where they have the option to call in or e-mail their requests. As a measure of preventing abuse, Amazon monitors repeated returns from users and freezes their accounts if it appears that they are taking advantage of the return privilege. For instance, if there is what Amazon deems to be an unusually high number of returns from a certain person, their account will be disabled. At the time of this writing, there is no posted return policy specifically for eBook returns, other than a directive on the FAQ page prompting customers to contact Amazon by e-mail in the event of technical problems, such as when someone is having trouble opening an eBook or when data is missing.

Advantages and Disadvantages of Electronic Publishing for Authors

Traditional publishing

Publishers come in all shapes and sizes. The field of publishing is diverse, perhaps due in part to the variety of age groups and subject matter it provides for. Not all publishers are created equally, and it is always in both the author's and the publisher's best interests for the publishing hopeful to thoroughly research and familiarize herself with the publisher's guidelines. While some publishers accept unsolicited (un-agented) manuscripts and queries, many more, especially in the larger houses, do not.

The mergers and acquisitions that have taken place in recent years — in which companies largely concentrated in New York City have merged to form corporate conglomerates — have resulted in the large publishing houses becoming even larger. With such moves comes increasingly limited accessibility for writers, who find that to gain acceptance from the publisher of their dreams, they must have an agent. But getting an agent to represent your work can be just as difficult as finding a publisher, and sometimes even more so.

Authors whose work is published by the large houses receive the benefit of widespread exposure from a company whose pocketbooks are deep enough to promote their already highly respected books. And where pocketbooks are concerned, most traditional publishers are concerned primarily with what will sell — what will help their bottom line. In the world of traditional publishing, unfortunately, many promising manuscripts are relegated to the slush pile and may remain there indefinitely.

Self-publishing does not hold the prestige of traditional publishing, and sometimes with good reason. After all, with traditional publishing, each manuscript must go through a guaranteed editorial process. Self-published works do not have such a stamp of approval on them, but occasionally a particularly motivated or talented author who self-publishes (and self-promotes) gains the attention of a traditional publisher. Once the contract is signed with a mainstream publisher, the potential to make considerable profit becomes more readily available.

Comparing Electronic Publishing

EPublishers are often small presses publishing in traditional print and online or electronic formats, or in electronic format only. They produce previously unpublished work, or reproduce in electronic format works that are currently in print or that have fallen out of print. Some ePublishers offer print-on-demand (POD) options, so users can print a hardcopy version. Due to the wide variety of ePublishers and the low production costs involved in ePublishing, authors may find their publication efforts more readily accepted by publishers through this avenue of publication, but should keep in mind that the benefits of sales, marketing, and prestige they might receive through traditional publication is greatly reduced in this market. Authors should use caution when considering ePublishers. Beware of subsidy or vanity publishers who market themselves as ePublishers.

The number and popularity of ePublishers reflects the changing face of publishing and the growing prominence of the Internet and computer-based technology in the publishing industry. In an era in which the online components of newspapers are gaining momentum — with some merging into Internet-only territory — and where cost-cutting, increased quality, and faster production are concerned, ePublishing makes sense. But, like other houses in the traditional, print publishing industry, not all ePublishers are created equally. Each may have — especially as time goes on and ePublishers become more established — its own particular tastes and preferences, along with greater competition due to increased demand for their services.

Enter Kindle. With large publishers like Random House, Penguin Group USA, and Simon & Schuster offering their publications as Kindle versions alongside small presses, independent publishers, and individual authors, the playing field is perhaps as close to level as it ever will be. Publications from large and small publishers, and from excellent and mediocre authors, are all available in one place. And the selections do not consist of just fiction — they are also magazines, newspapers, nonfiction, thrillers, children's literature, historic texts, out-of-print works, and many more in almost every conceivable genre. Add to that variety the fact that anyone can publish his or her masterpiece on the Kindle for almost nothing, and the nearness of the Kindle to the Internet in providing freedom of expression is clear.

Editing

The editing process is one that most careful writers incorporate into their writing and proofreading practices. If you are self-publishing on the Kindle, be sure to proofread and to perform a spell- and grammar-check. If you are not a strong speller or feel that your grammar and syntax could use a little help, enlist a pair of eyes (or several) to go over your work, in addition to the spelling and grammar checks. If you have the means, consider hiring a professional proofreader or editor to help finalize your work. Just because it is easy to publish on the Kindle does not mean you should gloss over the very important work of proofreading. You want your final product to be not only useful, but to generate a buzz and sell as many copies as possible; to do that, your manuscript must be professional.

The general editing process for traditional publishing may include all of the above, and then another round of editing with the publishing company's in-house editorial staff. After the book is in print, there is no going back and changing it. You should give the same attention and care to editing a book you intend to publish on the Kindle as one you intend for publication through a traditional market. The added benefit of the Kindle method of ePublishing is that you, the author, can go back and make changes after your book has been uploaded. Simply make the changes to the document on your computer, and then re-upload the file through Amazon.com's Digital Text Platform (DTP). Changes go through an approval process, and are usually reflected within a week. It is recommended that you e-mail DTP a notification of changes made, whether to the product details or to the content itself. Also notify customer service by clicking on the feedback link that appears on your title's detail page. Updated content replaces old content.

Marketing and promotion

When it comes to comparing marketing and promotional methods for traditional publishing against publishing your work on the Kindle — which is essentially self-publishing — the primary differences lie in the availability of financial resources and connections. A large publisher like Knopf Doubleday Publishing Group has an entire department devoted solely to marketing the books their company publishes. Their reach is far and wide, and can include distributors, libraries, bookstores nationwide, and then some. When Alan A. Example self-publishes his Kindle book, at best he may have a few helpful friends, family members, or perhaps a coauthor or two, who are willing to help him spread the good word.

For the most part, Alan A. Example will not have contact with distributors, libraries, or chain bookstores in multiple regions. He will have to rely on his own marketing and self-promotional methods to let people know his work is out there. Once Example sells several copies of his work, he may begin receiving reviews of his work on his book's Amazon detail page. Reviews and customer feedback can hurt or help a book's sales, as they are prominently featured on the book page. While you cannot remove them, clicking on one of the customer review links takes you to a page where you can report individual comments, if necessary. Customer feedback is a powerful marketing influence, but some entrepreneurially minded authors have tried various methods of advertising their work, visiting bookstores, meeting with distributors, seeking agents after self-publishing, and more. *These and other methods of self-promoting will be discussed in detail in Chapters 7 and 8.*

Electronically Publishing Through Amazon

Publishing a book for the Kindle is accomplished through Amazon.com's Digital Text Platform (DTP). The DTP is a publishing tool located at **http://dtp.amazon.com**, and is accessible once you have entered your username and password. If you do not yet have an Amazon account, you can sign up for free by clicking the "Sign Up" button on the right side of the page. Informational topics offered on the page include FAQs, general information about the Kindle, information on CreateSpace[SM] (a print-on-demand [POD] publisher that uses your electronic files to create a print version of your book), and the Amazon Community Forums. Users can access the DTP page from here or, for users who are familiar with HTML, the Mobigen Software and Mobipocket eBookbase, applications that are targeted for publishing on mobile devices, such as smartphones, PDAs, and tablet PCs. *For more on publishing for mobile devices, see Chapter 5. For a basic HTML guide, see Appendix B.*

Advantages

The ease of use, quick route to publication, and low cost involved in electronic publishing is appealing to many authors, especially those who have had little luck getting their work signed on with mainstream publishers. Many authors and publishers are beginning to see ePublishing as a forward-thinking strategy, rather than just a supplemental route to publication. As mentioned before, the nearly automatic widespread dissemination that is inherent in publishing electronically is a big draw for publishers and authors alike. Once a publication is available on the Internet and for the Kindle, it is accessible by potentially billions of users. Costs are reduced for buyers as well as publishers; most Kindle editions retail for less than $10.

Disadvantages

While an ePublished work is technically available to the masses, visibility is diminished in contrast to the widespread marketing agendas of large publishing houses. The Kindle is a platform that is constantly adding new titles; at the time of this writing, 725,000 titles were available, including newspapers, books, magazines, and blogs. With so many new titles being added all the time, how do you ensure that potential readers discover your work? Answer: The DTP allows you to add keywords to your book, which are search tags (words and phrases) associated with your title that help it to emerge in appropriate search results. *Keywords are discussed in further detail in Chapter 4.*

Adding keywords is not enough, however. Potential readers will want to see a preview of what your title has to offer, and it is important to make this available. A potential reader in a bookstore has the luxury of flipping through the book, and reading through it as much as they wish before deciding whether to purchase it; the potential reader on the Kindle must rely on the preview, title, cover, and book description to clue them in to your topic. Users are, however, provided with the option to download a sample before buying, which is typically one to two chapters.

Print-on-Demand (POD) Publishing and CreateSpace

Print-on-demand (POD) is a method of printing a single copy of a book on an as-needed basis; it is offered by some publishers as a convenient alternative to keeping stock in a warehouse, which may or may not sell. I discuss POD publishing primarily in the sense of its being an extension of creating an eBook; also, because it is something that Amazon offers to authors who publish their work through the DTP; and because it is simple to do if you have already created an electronic file of your book. It may be a natural next step for some people who choose to create eBook versions of their titles, and who wish to market their book through multiple avenues.

POD publishers are useful for printing in-print, out-of-print, and self-published titles, without having to keep stock in a warehouse or worry about unsold copies. Publishers can keep a book in print and maintain rights to the work. This option also allows libraries and anyone looking for a book no longer available through traditional means to access the work; publishers of academic and specialty books also find POD useful. An author utilizing a POD service for an unpublished title is essentially self-publishing.

Print-on-demand (POD) options are becoming increasingly common. Lightning Source (located online at **www.lightningsource.com**) is one POD option, and a division one of the largest book distributors, Ingram Book Company. Lightning Source typically prints and ships the book within 24 hours of the order. The author or publisher establishes the retail-selling price, including discounts for wholesale purchases, as well as a policy on returns. Lightning Source distributes through several major channels, such as Ingram, Amazon. com, Barnes & Noble℠, and Baker & Taylor, however, this means that your title will appear on a long list of other titles. The author's income is what remains after the cost of printing, and the fee for the service is $12 per year per title plus a one-time "minimal" set-up fee, according to the Web site.

Lightning Source offers a cover template generator, which is a pre-designed software template to help you generate a book cover, a spine width and book weight calculator, help with setting up book files, and guidance on avoiding common issues you might encounter with setup and file preparation. Note that the cover template generator requires that you first provide the design elements, and then set them in place within the template, which generates a (free) barcode. Normally bar codes are sold for a fee online through sites like Bowker.com. Lightning Source does not, however, provide assistance with editing, files, design, marketing, or promotion. To raise your title from the obscurity of the list in which it is buried, and to get bookstores to purchase it, you will have to do your own marketing.

Other POD options are Xlibris (**www.xlibris.com**), Author House (**www.authorhouse.com**), and iUniverse (**www.iuniverse.com**). Such traditional POD options can cost between $500 and $1,600 for initial set-up, and $15 to $50 and up per copy. Use discretion when considering any POD service, and be sure you have a complete understanding of all fees that will be required — as well as the services the company will and will not provide — before committing to anything.

A word of caution to authors with agents and those under contract with publishers: Find out whether there is anything in your contract about POD. Make sure the out-of-print clause in the contract contains a provision protecting your rights in the event that your book becomes available in POD format only. You will want to ensure that the rights to your work are not tied up indefinitely with a publisher who technically can always have a copy of your work available in POD format.

CreateSpace

Amazon has partnered with the POD publisher CreateSpace — good news for publishers of Kindle books. Unless you want to utilize their paid services, which include publishing, editing, marketing, and layout and design, it is free to use basic CreateSpace. CreateSpace publishes a variety of media, including books, video, MP3s, CDs, and DVDs. Located online at **www.CreateSpace.com**, CreateSpace is also a distributor, much like Mobipocket, who sells content through Amazon and other channels. *See Chapter 5 for a comprehensive discussion on Mobipocket.*

According to the CreateSpace Web site, the author-publisher agreement is non-exclusive to allow the author flexibility in their distribution and publishing choices. But as with any contract or agreement, read it thoroughly before signing, clicking OK, or otherwise accepting it. The POD model means that your book is not printed until an order is placed. Where royalties for books are concerned, CreateSpace takes 40 percent on each book sold on Amazon.com, plus a per-book charge, plus a per-page charge. The royalty for each book is different, and depends on the list price you set, the use of color in the book's interior, the size of the book, the number of pages, and the platform on which it is sold. To figure what royalties would be for your book, visit the CreateSpace royalty calculator, at **www.createspace.com/Products/Book/#content2**.

When you sign up for the program, you are provided with an ISBN or UPC free of charge. An ISBN (International Standard Book Number) is a number assigned for purposes of inventory and identification, and is issued to each title for shipping purposes; so all mass market paperback copies of *The Stand* issued by Signet in 1991 have the same ISBN, but any subsequent versions of the same title that are published — such as hardcover or expanded editions — have different ISBNs. The UPC bar code is used for merchandise. UPCs (Universal Product Code), when they appear on books, are used for mass-market paperbacks like those you may see in grocery stores and drugstores. UPCs cost more than ISBNs (about $300) and a single ISBN currently retails for $125 from **http://myidentifiers.com**. *For more on ISBN and UPC codes, see Chapter 2. This section is particularly helpful if you are not using CreateSpace or need a separate ISBN for your electronic publication.*

In early December 2009, CreateSpace announced a partnership with Lightning Source. Lightning Source Inc. is a division of Ingram Content Group Inc., the leading international book distributor. This new alliance means an increased distribution for CreateSpace Pro Plan members, which goes beyond Amazon to reach thousands of bookstores and online retailers, and even includes libraries — tapping the diverse distribution market of Ingram. This expanded option is only available for members who sign up for the Create Space Pro Plan, which involves a one-time fee of $39 for each book title; the renewal fee is $5 per year. The CreateSpace Standard Plan is free.

Considerations of Self-Publishing

As with any form of self publishing, be aware that, because of its very nature — that is, lacking the editorial procedures and quality assurance checkpoints of traditional publishing — you should consider carefully what you want out of self-publishing before venturing into it. If you are trying to build up solid credentials as an author, self-publishing may be the easiest but perhaps not the most professional way to go.

That being said, William Blake, Virginia Woolf, and Beatrix Potter were among several famous writers who developed and printed their own materials, in part because it was a means by which they could achieve their personal visions for their publications. Teachers and professors have been known to create spiral-bound versions of self-developed instructional materials that serve their purposes of limited distribution to their students. Family members might want to distribute a cookbook of their own creation, or perhaps a book takes the form of a family memoir or well-loved bedtime story made up for your children. Other reasons to self-publish might be to have full control over your work, to maintain all rights, or to see your work in print when you are having difficulty being published by traditional means. Keep in mind that self-published nonfiction tends to fare better in the marketplace than the fiction alternative, largely because it represents a niche or specialized market.

Some writers have experienced success via widespread distribution of their self-published work, but they have had to be clever, work very hard to promote their books, and have had a bit of luck on their side. The more successful authors in this group (who include Richard Paul Evans, author of

The Christmas Box, and Christopher Paolini, author of the fantasy series *Eragon*) established a successful sales record and had their work picked up by a traditional publisher. The works then went through the industry's largest and most widespread marketing and distribution channels, after which these authors' pocketbooks and reputations grew. They were exceptions who started out self-publishing, and when their work was successful, they entered the world of mainstream publication.

Reality Check:
Setting Realistic Expectations

Honestly assess your needs and what you want out of publication. Understand that, whether you publish traditionally or self-publish, financially speaking, you are not likely to be the next J. K. Rowling. Aim high and always do your absolute best work, but know that, realistically, you will probably not get rich with self-publishing. If you want to publish a book just to make money, you should read further in this book to gain a better understanding of royalties, rights, advances, and the manner in which authors are paid. If you do begin to see a steady stream of income from your writing — whether through books, or a combination of books with articles, speaking engagements, tours, and related methods — only then should you think about quitting your day job. It is possible, but requires unwavering passion, commitment, and dedication.

While it is true that you get to keep a larger percentage of royalties with self-publishing than you would with traditional publishing, consider that your audience is also much more limited, and therefore so is your profit margin. Realistically, there are other things to consider about self-publishing. While the quality of self-published work has increased due to advancing technology — lacking an editorial process, the contributions of a design team, a national or international distribution market, and the recognition of industry professionals — it is not often taken seriously. Authors who are published through traditional means work hard to perfect their craft and their manuscripts. The books you see on library shelves may well have started life as a manuscript that had many doors closed on it, went back to the author's desk for revision, and was then sent out again to dozens of publishers before being picked up by one.

Some people will tell you that it is next to impossible to be published in the trade and mass market industry, but you should ask yourself how all those books made it to their places on the shelves of your local library and bookstore. Traditional publication is far from easy; yes, the numbers may be against you, but if you are serious about your craft and work hard at it to make it the best that it can be, it will be well worth the effort it takes to achieve it.

Later in this book, you will read about multiple ways to cross-promote and market your Kindle publication. Do not prepare to publish a title for the Kindle expecting instant wealth, blockbuster success, and a major-league contract offer. Be prepared to put in a lot of hard work — writing the book, editing and proofreading it, planning the layout and design, publishing it (and proofreading again) on the DTP, and then marketing and promoting your title. If you do all these things, do them consistently, and maintain patience and persistence, you can prepare to reap the rewards of your *efforts. Chapter 7 focuses on marketing and promotion, and Chapter 8 discusses how to make use of the Internet and social media to promote your work.*

Kindle Competition

There are currently more than two dozen eBook readers on the market, manufactured by more than a dozen companies. Some of the leading Kindle competitors are:

- Barnes & Noble nook™
- Sony® Reader™
- iriver Story
- Bookeen Cybook Opus
- Hanvon™ E-book readers
- BeBook℠ eReader
- Apple™ iPad™

Sony's line of eBook readers includes the Pocket Edition™, the Touch Edition™, and the Daily Edition™. Their product line is viewable online at **www.sonystyle.com**. According to the Sony Web site, the readers allow access to more than one million free public domain titles from Google Books. The nook from Barnes & Noble is one of the newest additions to the market, and along with the Sony, appears to pose the biggest competition to the Kindle. The Kindle, however, is arguably the most widely marketed product of its kind, and the one with the highest visibility. The nook, the iriver Story, the Kindle, and most of the Sony readers feature touchscreen technology, but vary in terms of how many books they can hold, battery life, supported file formats, the presence or absence of a qwerty keyboard, and other considerations, such as the type of wireless connectivity offered.

According to a WirelessWeek.com article from Nov. 30, 2009, the Kindle has been the longtime leader of wireless reading devices, and November 2009 constituted its best-of-all-time sales month. The Kindle remains Amazon.com's No. 1 bestselling product. The nook, which debuted Oct. 20, 2009, is up against stiff competition. However, according to the Wireless Week article, the nook was back-ordered as of the end of November 2009. The Apple iPad was released April 3, 2010, but was introduced to the public on Jan. 27, 2010. It features a 9.7-inch, LED backlit screen (as opposed to the Kindle's use of E Ink® to avoid backlighting). The debut price started at $499.

Wireless reading devices with the magnitude of the Kindle possess the technology and platform to revolutionize publishing as readers, writers, and publishers know it — and some may argue are well on their way to doing so. Since the advent of the Internet, publishing and communication have changed irrevocably. Electronic publishing is still in its fledgling stage, and many eyes are watching to see where it goes.

WHAT IS THIS MYSTERIOUS THING KNOWN AS E INK?

Simply put, E Ink does not use backlighting. Backlighting is the internal lighting used in computer monitors and other devices that essentially illuminates the pixels on the screen from within the device. E Ink technology uses positively and negatively charged black and white plastic particles that are rearranged according to what should be on the screen. How does this happen? Basically, the particles are coated in oil, which allows them to rotate freely within a transparent sheet of silicon. Think of it like a sophisticated Etch A Sketch®.

Setting Up an Account Using the Amazon Web Page

This chapter assumes that you have an Internet connection, a Web browser such as Mozilla Firefox or Internet Explorer, a working e-mail address, and a book manuscript in one of the following electronic formats: HTML, .doc, .docx, or .rtf.

Registration process

If you do not already have an Amazon account, visit **www.amazon.com** to create one. On the homepage, scroll down and look for the sidebar on the left of the page titled, "Features & Services." Under the subheading, "Selling with Amazon," click "Publish on Kindle." You are taken to a page titled "Amazon Kindle's Publishing Program," where you will click the Digital Text Platform (DTP) link. You are taken to the DTP page and prompted to sign in through the box on the right. New users click the "Sign Up" button in the box below and fill

CHAPTER 2:
Getting Started: Amazon and the Digital Text Platform

out the form when prompted to do so. Follow the remaining registration prompts.

Setting Up an Account Using the Digital Text Platform Web Page

Once you are signed up with Amazon, navigate to the DTP Web page (as described above) and sign in with your e-mail address and password. You will be asked to agree to Amazon's digital books program terms and conditions, and will be taken to the DTP, where you will see three tabs at the top: "My Shelf," "My Reports," and "My Account." *For more information on the terms and conditions, see the section "Amazon's Security Measure" in Chapter 4.*

"My Account" page

In the DTP, navigate to the My Account page by clicking the appropriate tab at the top of the page. Here, you will enter the following information to ensure you receive royalties:

Company/Publisher information

Under Company/Publisher information, enter your name in the Full Name/Company Name field if you are an individual who is self-publishing. If you are a publisher or have a self-publishing business set up where you publish your own titles, enter the business name. If you are uploading a book released under a publisher, you must own the copyright to the book before uploading it. Check your contract if you are unsure. If a publisher owns the book's copyright, be sure to check with them before proceeding further. You may not have the right to upload your book. Enter the appropriate address.

Payment information

You may choose to be paid via electronic funds transfer or by check. Note that to have payments transferred to you electronically, you must have a U.S. bank account. All others will receive payment via a check in the mail.

Selecting the electronic funds transfer option will prompt you to enter information in the following fields:

Business Type

Under Business Type, you may select from three different business types: Individual, Partnership, and Corporation. Select "Individual" if your book cannot be affiliated with any business. Choose the status that best reflects your legal standing and the book's relationship to it.

Social security number, tax identification number, or employer identification number

If you selected "Individual," enter your social security number. Business representatives will enter a tax identification number (TIN) or employer identification number (EIN). Partnerships, employers, and corporations primarily use TINs and EINs. It is a number that identifies any business entity. Applying for a number is a free service provided by the Internal Revenue Service. See **www.irs.gov** for details.

The purpose of entering the SSN, TIN, or EIN is so your income can be reported to the IRS.

Bank account

To enter your bank account information, click the "Add or Edit account" button. You will be prompted to fill in your bank name, routing number, and account number. Refer to your checkbook or online checking account to ensure accuracy of all information. Click the "Save entries" button to save your information. A box appears prompting you to wait while the information is saved.

Overview of Digital Text Platform

Amazon offers a DTP Quickstart Guide at **http://forums. digitaltextplatform.com/dtpforums/index.jspa**. Under the heading, "Browse the Knowledge Base," click on "Getting Started Guide" and then "Amazon DTP Quickstart Guide." The link takes you to a PDF version of the most recently updated guide. Updates to the guide should be posted to keep users aware of Kindle software upgrades.

Open beta version of publishing software

On the DTP homepage, if you look in the top left corner, you will notice that the Digital Text Platform logo has the word "beta" situated in superscript just to the right of it. This indicates that the DTP is still in beta — or testing — format. In other words, DTP is still in a stage of operational development, and, according to techies, the DTP system is a complex and unwieldy beast, not yet stabilized. DTP is a wonderful system; it simply does not have all the inconsistencies worked out just yet (that is, at the time of this writing).

Community support forum

To visit the DTP Community Support page, go to **http:// forums.digitaltextplatform.com/dtpforums/index.jspa** to visit the public forums, which is a common area for users to post and answer questions. The "Browse the Knowledge Base" header provides information, such as help for creating an account, publishing content, formatting in HTML payment information, and utilizing the dashboard (the main DTP page). The "Discuss in the Forums" header provides access to a FAQ page, support for publishers, and the community forums. *For a basic HTML guide, see Appendix B.*

Drawbacks

Inadequate technical help

The DTP forum is available for users to post their questions, but eliciting responses from administration is achieved rarely. There appears to be no dedicated, systematic response functionality. The forums often reflect the sense of frustration users feel at problems that occur. A post from "DTPadmin" dated Feb. 9, 2009 reads:

"Currently, there are two ways to get support for Amazon's Digital Text Platform:

1.) Browse the list of Frequently Asked Questions in this forum to make sure your question hasn't been asked before. You should also check the 'Ask the Community' forum.

2.) If you don't find the answers in the forums send an email to **dtp-feedback@amazon.com.**

These are the only two ways to obtain support. Please do not send email to any other amazon.com email address, as this may delay your response.

Thanks for using Amazon's Digital Text Platform."

Apparently, responses are posted from the engineers who created the DTP, as they are the personnel who understand the system most thoroughly. There does not appear to be, at this time, a staff dedicated solely to handling issues and troubleshooting.

However, the Web site eBook Architects (**www.ebookarchitects.com**) can prove useful if you run into difficulties with formatting or need extra technical assistance (be aware that they charge a fee for their conversion services; rates vary and you should check the Web site for the most up-to-date prices). Smashwords™ (**www.smashwords.com**) is also helpful; you can use the service to convert a .doc file for distribution through Smashwords, Amazon, and other channels; authors receive 85 percent of a title's net proceeds.

Benefits

Access to market

One search through Amazon's print listings provides ample evidence of the availability of a multitude of titles, including selections from new, emerging, and established authors alike. All have equal access to the same market. Unlike the disparity among titles published from houses as dissonant as, say, Random House, Inc. and Tachyon Publications, a small publisher of science fiction and fantasy has equal access due to the Kindle, and more broadly, the Internet. Any publisher can potentially access a market consisting of readers searching for something in particular, even if that something is only something good to read. When you walk into a bookstore, you go in knowing that only a certain number of titles from certain publishers are represented. When most people look for something from an independent publisher or author, they go to the author or publisher's Web site to place their order — not the bookstore. Kindle provides a marketplace where all are present and available for purchase. The embedding of relevant keywords behind the scenes (in the code) for your book ensures that your title appears in the appropriate search results.

Demand for independent authors

As an independent author — that is, one not linked with a publisher — you make your work available to the masses. With the growing popularity of the Kindle, there may come an increased appreciation for independent authors, although the change may occur gradually. Whatever the case, independent authors and self-published authors are on technically equal footing with works published through mainstream houses; all they need to do is market their titles to sell.

Access to successful independent publishers

Much like readers have equal access to independently published authors, they also have access to independent publishers. The DTP platform is available for use by both individual authors and publishers, large or small. Currently, a search for literary fiction turns up results primarily from larger publishers like HarperCollins and Simon & Schuster, but as Kindle editions catch on, undoubtedly more small and independent publishers will begin to funnel their fare through what could be a potentially lucrative market, while at the same time satisfying the tastes of the particular readership that seeks such publications.

ISBNs and UPC/Bar Codes

Before publishing your book, you may want to look into obtaining an International Standard Book Number (ISBN). The ISBN is a measure of inventory and identification that is issued to each title for shipping purposes. Different editions of the same book receive different ISBNs — such as the print and electronic versions of a publication. For example, all copies of Margaret Atwood's paperback edition of *Oryx and Crake*, printed in 2004 by Anchor, carry the same ISBN; however the hardcover edition released in 2003 by Bloomsbury Publishing PLC, carries a different ISBN, which is attached to all copies of that edition. An ISBN is acquired through **www. isbn.org**, where you will also find price listings. The numbers are available for purchase in blocks of 10, 100, and 1,000. The ISBN Web site posts a recommendation that publishers purchase enough ISBNs to last them for five years. This is for a few reasons. The more numbers purchased, the less they cost. You will also be able to maintain the same publisher prefix for a longer period of time. If you require only a single ISBN, they are offered at **http://myidentifiers.com**.

If you have ever looked at an ISBN, you may have noticed that they come in a few different forms: ISBN-10 and ISBN-13. This signifies a move from 10-digit to 13-digit numbers to accommodate the growing number of titles. Additionally, the current prefix "978" will soon move to 979. According to the ISBN Web site, the "X" that appears at the end of the number stands in for the number 10; the last number is referred to as a "check digit." The ISBN-13 has five parts: the prefix, group or country identifier, publisher identifier, title or edition identifier, and the check digit. The ISBN-13 is now the industry-recommended preference.

The bar code represents a translation of the ISBN. Authors or publishers can request bar codes from the ISBN Web site or from **www.bowkerbarcode.com**. Why invest in a bar code? Most retailers use the bar code for scanning during checkout and sometimes for inventory, and most retailers and wholesalers will not carry your book without a bar code. When selecting your bar code, choose the Bookland EAN/13 with add on. This should appear on the lower half of the back cover, which is known as "cover four," on both paperback and hardback editions. The bar code you will be using differs from a UPC bar code, in that the UPC bar code is used for merchandise. UPCs, when they appear on books, are used for mass-market paperbacks like those you may see in grocery stores and drugstores. UPCs cost about $300, and it is not likely that this is the type of bar code you will require. The cost for ISBN/bar codes varies depending on the quantity you wish to purchase, but ranges from $245 to $1695. Access the order form for block ISBN purchasing at **https:// commerce.bowker.com/standards/cgi-bin/isbn.asp**.

Chapter Conclusion

Signing up and getting started in DTP is a fairly straightforward process. While the DTP program is still in the beta testing stage, it is a powerful program, and will be even more so when the bugs have been worked out. Buyer beware: Technical support is very limited at this time, as there is no dedicated support staff to handle troubleshooting and inquiries. An e-mail sent requesting help may take an indeterminate amount of time to get a response. Outside of such difficulties, the DTP is an easy program to use, with the result being that of publication — a reward and an opportunity in and of itself. The ability for independent authors and presses and large publishers to exist within the same realm, or the same virtual storefront, is one that not only expands the choices available for the readership at large, but also opens doors for individuals and publishers whose pre-Kindle options for generating a revenue stream have been all too limited.

CASE STUDY: EMBRACING CHANGE IN THE PUBLISHING INDUSTRY

Brett Rosenblatt
Software Engineer and
Entrepreneur, F-Line Technologies
brettgarciarose@gmail.com

Brett Rosenblatt is a software entrepreneur and writer living in New York City. After graduating from Northeastern University with a degree in journalism and creative writing, he worked as a freelance journalist for Boston and New York publications, as well as an investigator for various organizations in the animal rights and environmental movements. After returning to NYC to study at Columbia University, he founded F-Line Technologies, inventing and producing software for the global telecommunications industry. Brett's fiction can be found in various literary and consumer publications, including Newsday Magazine, Rose and Thorn, Opium, Withersin, Sussurus, Metazen, Spectrum Magazine, and others.

As one of the simplest forms of art and entertainment, it is curious that literature has dodged the technological revolutions that have transformed other mediums over the past several decades. The media card in my camera holds thousands of photos. Yet, in my bag, I can carry three books at most. We like "curling up with a good book." But is this the best way to ingest content? Can this method be improved? Kindle says yes. For your ten-day trip to the beach, would you stop at the library and check out 20 pounds worth of books? Probably not.

What is interesting about the eBook revolution is that it offers the very real possibility that everyone can become an author. The Kindle represents a significant breakdown of the publishing world's gatekeeper mentality. In traditional publishing, if there were no barriers to being published, there would be no industry. There is only so much space, only so much paper, only so many bookstores, and only so much time to read. Is the Kindle changing some of those things? Yes. But the Kindle will not edit prose. It will not correct grammar or enliven dull stories. Literature will not type itself. Technology does not corrupt art; business and money do.

All that changes is the delivery method — however, that changes everything. Because, as history so eloquently and repeatedly teaches us, there is no way to contain art. As a writer, my passion is to communicate passion. My business and life is to communicate. I simply have more options: to do it through an eBook, through agents and publishing houses, through newspapers, indie magazines, or even through Facebook℠, blogging, or Twitter®.

How Amazon Pays Authors

Royalty rates and royalties

Royalties are earnings an author makes from the sale of books — a percentage of the list price, which you set before publishing your work for the Kindle. Amazon allows you to choose a price for your title between 99 cents and $200. You continue to earn royalties while your book is listed with Amazon. In traditional publishing, a royalty agreement is bound into a contract and the method and frequency of payment is handled differently.

Amazon pays authors a certain percentage of the profit for every Kindle eBook that sells. Amazon makes a profit from your work, but so do you — 35 percent. If you feel uncertain about allowing Amazon to take a 65 percent cut of your sales, consider that you are not self-publishing a print version of your book, in which case you would have to purchase ISBNs and hire someone to do your cover art, book design, editing and proofreading, typesetting, and pay for the costs of printing in addition to that. You would still be responsible for marketing and promotion, and so

CHAPTER 3:
Cashing In:
The Amazon Payment Process

much more. Amazon automatically assigns their version of an ISBN number — the ASIN — to items in the Kindle Store and other merchandise that does not carry an ISBN. You can certainly purchase an ISBN for eBooks, but it is not required. Obviously, you should still market and promote an eBook just like you would a book in print, but the demands for self-publishing an eBook are much less financially burdensome — not to mention less demanding of your time — than they would be if you were self-publishing a print title.

When publishing for the Kindle, you also do not have to deal with agents or publishers, who might not give you the time of day if you are a new or inexperienced author whose writing is not quite up to snuff. If you write articles, you have even more reason to want to make a profit from your work, and if you have ever tried to earn a living from submitting your articles to magazines, journals, or newspapers, you know how difficult that goal is to reach for all but the most driven and talented of writers.

All things considered, Amazon's 65 percent profit is not so bad. They developed the Kindle and the platform that allows authors to publish their work free of charge and promote it along with all the other titles from very well known authors to large corporate publishers. You are paying partly for the promotion that is inherent in being a part of such a high-profile marketplace. While it is mainly up to you as an author to make sure people know your book exists and how it can help them or meet their needs, just being there is half the battle, and Amazon provides that opportunity.

Prior to the launch of the much-discussed Apple iPad, a competitor eReading device, Amazon announced a new option in their royalty plan. In a press release dated Jan. 20, 2010, Amazon announced that they plan to provide authors with the option to earn 70 percent of royalties on their Kindle eBooks, instead of the original 35 percent. According to the press release, the new scheme is slated to make its debut on June 30, 2010, and will be offered as an option in addition to the standard 35 percent royalty plan, rather than in replacement of it. However, some caveats apply. To qualify for the new plan, your eBook must meet the following requirements:

- The list price supplied by the author or publisher is between $2.99 and $9.99

- The list price is at least 20 percent lower than the print book's lowest price

- The title is available for sale in all locations in which the author or publisher has legal rights

- The title will enable text-to-speech

- The work must be in-copyright and not public domain

- The book must be initially sold in the United States

HOW MUCH ARE THE ROYALTIES?

Traditional publishers' royalty rates are standard at 10 percent. Amazon pays their authors 35 percent of the list price, or retail price, which you set. Amazon often discounts books to encourage sales, but you still make the same. If you set your list price for your Kindle book at $9.98, your profit is $3.49 for every copy you sell. If Amazon decides to mark down your book to boost sales, and offers it at a discounted rate of $7.49, you still make $3.49 for every copy sold.

WHEN ARE ROYALTIES PAID?

Payments are sent from Amazon 60 days after the close of the calendar month in which sales occurred. You do not have to provide an invoice to Amazon; rather, when you accrue a balance of $10 or more, your payment will be sent to the bank account you provide during set-up in the DTP.

HOW ARE ROYALTIES PAID?

Payments are made via electronic funds transfer (EFT) to the bank account you entered when you signed up in the DTP. To change this information, simply log in to the DTP at **http://dtp.amazon.com** and click on the "My Account" tab.

Digital Text Platform
My Reports Page

At the main DTP page, located at **http://dtp.amazon.com**, click on the "My Reports" tab at the top. Under "View Reports," you are provided with several options:

- **View Month-to-Date Report:**
  Allows you to view the current month's summary of sales transactions.

- **View Previous Months' and Year to Date (YTD) Reports:**
  Includes year to date sales transactions, with a view of sales summaries from previous months. Summaries are released on the fourth of the month. You will use

these when you file taxes, as Amazon does not remit W2s at the time of this writing.

Sales ranking vs. actual copies sold

A note in the "My Reports" section informs users that the reporting of sales ranking does not reflect sales of their individual titles, but rather is interdependent with the sales of other books. The note reads: "Sales ranks for your books are updated regularly throughout each day, but are not updated automatically with each sale." The note further warns that "sales ranks are relative" with those of other authors. At one time, there was quite an upheaval in the forums. DTP users' questions and posts reflected the frustration they felt with the inaccurate portrayal of their sales rankings. Many felt that, since their sales rank showed improvement, it meant they were selling copies. When those apparent additional sales did not show up in their sales reports, they were understandably frustrated at their perceived loss of royalties — a frustration that was only compounded by the inadequate technical support available in the forums.

Royalty Rates and Services of Competing Digital Publishers

To publish your work to the competition, especially Barnes & Noble's nook and the collection of Sony Readers, visit **www. smashwords.com**. This site allows you to join free of charge, to create an account, and upload your text. Smashwords converts uploaded text into multiple formats, and makes it deliverable to the iPhone, the Kindle, and its competitors.

The books are also made available for purchase on the Web site. Smashwords takes a commission of 15 percent on each book sold, unless you offer your eBook for free, in which case Smashwords charges nothing.

Authors who publish with Smashwords retain full rights to their work, maintain control over which formats their work is converted into, and set the amount of sample text that will be seen by users. Smashwords is useful for people who do not have access to a U.S. bank account, as they pay their authors through PayPal℠ (**www.paypal.com**). Amazon made a change to the Digital Text Platform on Jan. 15, 2010 that removed the obligation for publishers to have a U.S. bank account, thereby opening up the program to international users. Smashwords uses a variety of quality control checks before sending titles — via their Premium Catalog — through mainstream channels, such as Barnes & Noble, Sony, and Shortcovers, but anyone may submit their work to the site. Through the Smashwords partner, **www.word-clay.com**, authors can select to have their books published in print form.

Another Web site, **www.booklocker.com**, offers a service similar to Smashwords, but Booklocker screens submitted works for quality control and commercial appeal. This service offers a print service for an additional fee.

Amazon.com takes a 65 percent commission on books sold through the Kindle Store; the Barnes & Noble Web site provides no information on whether they take a commission on eBooks sold to the nook.

Royalties in the Traditional Publishing Industry

There may be reputable ePublishers who will take your work through the traditional process of submission, consideration, acceptance, contracts, editing, and publication; and you should consider using these. While an ePublisher may not pay an advance, their royalties may pay out higher, often from 24 to 75 percent of the selling price. However, if you are simply looking to make your work available and do not wish to have others pay a fee to access it, you may consider another avenue, like desktop publishing.

Some publishers who promote themselves as ePublishers are actually vanity publishers or subsidy publishers. These types of publishers print books at the author's expense. If the publisher requires any sort of fee, it is in the best interests of your career and reputation to avoid them. Such companies may ask for a monthly fee, or a maintenance or set-up fee, and make most of their money from authors in this manner, rather than from sales of the book.

While it is true that you get to keep a larger percentage of royalties with self-publishing than you would with traditional publishing, consider that your audience is also more limited and therefore so is your profit margin. However, with Kindle sales on the rise and thousands of new titles being added to the Kindle Store on a monthly basis, the audience for ePublishers is rapidly expanding.

Thinking of royalties, the following are the main types of contracts you will see in the traditional publishing market:

Types of contracts

Work-for-hire contract: A flat fee is paid for work to be done and no royalties are offered. Most contracts give exclusive rights to the publishing company, and the work is copyrighted in the publisher's name, rather than that of the author, even though the author's name may appear on the book cover. Sometimes provisions may be made for additional payments to the author according to copies selling past a certain number, or when the author helps to sell copies of his work.

Royalty contract: While the advance varies depending on the book's marketability and the author's experience, royalties are standard at 10 percent. If the book is illustrated, the 10 percent is divided between author and illustrator; if the author and illustrator are the same, she will receive the entire 10 percent. If the book is not illustrated, the author collects the 10 percent. Both royalty rates and advances can increase (through an escalation clause) along with your sales and experience. Be sure to find out how royalties are calculated for foreign sales.

Clauses are portions of a contract that set out specific rules and regulations pertaining to all aspects of a book's publication. The following are a few contract clauses that affect royalties:

Net proceeds clause: This type of clause allows for royalty payments of net proceeds rather than gross. If your books are purchased at a discount of 30 percent, that 30 percent discount cuts into your earnings.

Sometimes a publisher offers royalties at a higher rate to make up for the loss in earnings.

Reserve clause: A certain percentage of royalties, or a "reserve against returns," is held by the publisher to cover the return of any unsold copies of your book by bookstores. The clause will specify a certain percentage of sales to be withheld and the duration the money will be held. A suitable time period to allow for the reimbursement to you of this money might be about a year.

Payment in the Print Market

Advances are usually paid to the author in two stages: the first half paid upon signing the contract, and the second half paid on the book's completion. Advances vary depending on the project, publisher, and experience of the author, but whatever advance you are given must be paid back in sales before royalties can be collected. Royalties are percentages earned on each copy of a work sold. Royalties can be "list" or "net." List provides for a percentage of royalties to be paid based on the cover, or list, price. Net is more often employed by specialty publishers or small and independent publishers who base the percentage on the actual profit made from the book. The net profit may take a cut if the book sells at a discount, and net receipts are sometimes referred to as "amount received" or "net price." List is based on the retail price, and net on the net receipts. During contract negotiations, you would need to make sure that your royalty is not reduced for smaller print runs.

The contract should include a provision that protects you from having to pay back the remainder of an advance not paid for by sales. This is not an uncommon scenario; books frequently go out of print before the advance can be paid off. You can increase your sales dramatically by helping to market your own work and with what is probably the single most effective marketing tool — book reviews.

Selling reprint rights to your book earns you additional money, and often the money earned from selling the rights is split exactly in half with the publisher. It can often take several years worth of sales to pay back an advance, so that initial $5,000 (or whatever you are paid) is an income spread out over that several-year period. To earn a steady income from a book, you would need to be an established author or have a book that is a true marketplace success. But the more books you have published, the more advances you will earn and the greater your potential to sell reprint rights or sell enough to start earning royalties.

CASE STUDY: CHILDREN'S BOOK AUTHOR DISCUSSES EBOOK IMPROVEMENT

Nik Perring, children's book author
E-mail: nikperringwriter@gmail.com
Blog: nikperring.blogspot.com
Web site: www.nperring.com

CLASSIFIED CASE STUDIES™
directly from the experts

Nik Perring is a writer and workshop leader from the UK. His short stories have been widely published in places including Smokelong Quartlery, 3 AM, Metazen, Ballista, Word Riot and Dogmatika. He is the author of the children's book, I Met a Roman Last Night, What Did You Do? (Educational Printing Services Ltd., Sept. 2006).

To be honest, the process of getting my book published in the mainstream print market was pretty straightforward. I had an idea for a book for children — historical fiction, based on the National Curriculum (UK), which would be educational. After I'd finished it and felt it was good enough to be "out there," I started looking for a publisher whom I thought would be interested, and happened upon Educational Printing Services, Ltd. (they specialize in educational books). EPS was the only publisher I approached, and they liked what I did, thought they could sell it, and offered me a contract very quickly. There were edits to do and bits that needed rewriting, which all went smoothly. About a year after I'd initially approached them, my book (*I Met a Roman Last Night, What Did You Do?*) was published.

I like the idea of eBooks. I think they're going to be a considerable part of publishing's future. But they're not quite there yet — I think they've got quite a distance to go before they begin to reach toward their potential and usefulness.

For them to get closer to that potential, I think three things need addressing:

First, the electronic readers, the devices the eBooks are read on, need to be a good deal cheaper. I think it's fair to say that a significant majority of those who have bought electronic readers are both book and gadget enthusiasts — that's not a sustainable market, nor is it one that's going to grow too much, especially with the limited choices of content on offer.

Second, the titles themselves need to be cheaper. There's much to be said for having the ability to carry around with you a whole library of books, but that becomes less appealing when the electronic versions of those books are almost as expensive as the physical versions. As I see it, electronic readers should be practical companions to physical books — the two should exist side by side and should not, on the whole, be competitors. That means that I, as a consumer, might want to own both the electronic and physical versions of certain titles (so I can take old favorites on holiday, for example, without the extra weight and bulk). If I've already bought one at, for argument's sake, $14 (recommended retail price) I wouldn't want to pay a similar price for the electronic version, simply for the convenience of having it on my electronic reader (which I'll have already forked out a couple of hundred dollars on). One way around this — and an idea I'm fond of — would be to offer people who had already purchased one version of a title the opportunity to buy the other at a reduced rate.

Third, availability and choice need to be improved. For this whole electronic thing to work I, as a consumer, with varied and eclectic tastes, need to have access to pretty much everything I have access to via bookshops (on and offline) if not more. I'm primarily a short story writer and was disappointed to find that Lorrie Moore, one of the most important and popular short story writers of the past three decades, had only one book (a novel) available from the Kindle Store when I searched for it on my iPhone's Kindle app (January 2010). There also needs to be one form of eBook file — so I can buy it from whichever retailer I choose and be able to download it onto whichever electronic reader I happen to own. And that's without mentioning the potential for out of print books and individual short stories and poems. Quite simply, the more stuff that is available, the more potential there is for sales, and the more reasons people will have for wanting to join the 'electronic book club.'

I think publishers, manufacturers, and retailers need to look at the iTunes/MP3 model, as opposed to VHS and Betamax. If I have an iPod, I can transfer what I've already purchased to it without buying the whole album or song again. If I want to buy something new, I have a considerable choice of where I can buy that from — a choice that's not dictated, or limited, by the sort of MP3 player I own. I think that's the scenario the electronic book world should be pushing toward.

However, do I see any benefit to publishing an eBook version of my existing title? No, I don't (not that I'd complain if my publishers wanted to put out an electronic version — I'm just not convinced there's a market for that particular book in that particular market, just yet). I think that electronic readers are, at the moment, something that adults, rather than children, will own (because of their price) so I don't see how I or my children's book would benefit.

What I do think though is that there's an enormous potential market for this in the future, especially with educational books. If the electronic readers are affordable and the books interactive, with

links to encyclopedias, course notes, textbooks, and the like —
they could be of huge benefit to those in education and could even
change the way things are taught. We've already seen how well the
iPhone has worked in classrooms, after all.

So, yes, I like the idea of eBooks and I think there's much to look
forward to in the future; they have enormous potential to add to the
world of literature and I'm looking forward to that potential being
realized, or at least being approached.

Further Examination of the Traditional Publishing Industry: Literary Agents

If your Kindle publication will be your first, or even if you have published before, it can be helpful to have an understanding of the ways of the traditional print world. Since you may be dealing with an agent at some point in your career, especially if you plan to continue down the path of publication, you should inform yourself as much as possible on how agents work. Almost every large publisher will only read work that comes through an agent — which may be good to know if your Kindle publication ever hits the bestseller list and you hope to someday have a publishing contract in your sights.

Agents serve as informed and well-connected liaisons between publisher and author in the world of traditional print publishing. Their job is to keep abreast of changes in the publishing industry, such as when an editor changes houses, or regarding mergers and acquisitions that take place. Agents are connected with editors throughout the publishing industry, and they have an understanding of what imprint or house is currently publishing what type of book, who they are publishing, and what they are looking for next. Editors are usually only a phone call away for well-connected agents.

Agents also prepare the submission package. They are helpful in this regard, since they are already familiar with the publisher's guidelines, and know who wants a query letter, who will accept a partial or a full manuscript for review, who prefers hard copy format, and who accepts electronic correspondence. When it comes to contracts, agents help negotiate the best deal for you. Keep in mind that the more money you make, the more the agent makes, so they will be looking for the best deal from a publisher. Experienced agents can detect weaknesses or flaws in a contract or deal. They are also useful for providing insider advice, and can advise when something is industry standard versus a bit out of the ordinary.

Many larger houses will only review new material that comes through an agent. This is not snobbery, but rather a measure of standardization that helps business run much more smoothly. Agented submissions cut down significantly on the materials that editors and readers have to go through to find what is suitable for their house. Good agents know what a publisher wants and what he is looking for, and will send them material accordingly.

You most likely need an agent if you have a fiction manuscript and want to be published by a larger or mid-sized house. Many agents will only represent authors who have already proven their marketability; in effect, those who have at least one successful publication to their name. However, it is possible, though rare, to attract a publisher's attention with a query letter. If your letter presents a brief idea of a story that suits their house and is something they would be interested in publishing, they may request to see a partial of your manuscript, which is then no longer unsolicited.

Agents usually collect 10 to 15 percent of an advance paid to the author, and 10 to 15 percent of the royalties. Artist's reps (an artist or illustrator's equivalent of a literary agent) charge anywhere from 20 to 35 percent. If foreign rights or movie rights are involved, the percentage will be higher, depending on the deal. Agents who collect fees may be unqualified for their positions or can be rip-off artists. Some agents charge authors fees as a source of income when they are not well-connected enough to make a living from commissions. However, agents commonly charge fees to compensate for legitimate costs that are out of the ordinary, such as the retyping of a manuscript, overnight delivery, long distance and out of country calls, and copying beyond what is average. Such costs will be made clear up front. However, keep in mind that agents are not editors, and if an agent offers editing services or makes promises to get your work into a certain house for additional fees, they may well be guilty of disreputable practice.

Protecting Profit: Self-Employment, Taxes, and Accounting

Freelancing versus traditional employment: business and taxes

If you are publishing your work for the Kindle and earning money from it, you are required to report your income to the Internal Revenue Service. If you earn income in this manner, there are some things you need to be aware of. When you work for someone else, he or she handles your taxes. If you do freelance work, you are required to report your income on your yearly taxes. Keeping good records of work done is not just important for purposes of tax reporting, it is also important to determine the amount of time you are putting into the work and the money you have to show for it. At first, this accounting might be confusing (and even disheartening as you look at the numbers), but keep in mind that successful freelance work takes time, both to build a client base and to build your reputation as an author.

Here are some basics to keep in mind when it comes to taxes and business record-keeping:

- Keep personal and business records separate, even if it means setting up separate bank accounts.

- Keep proof of all business-related expenditures, such as receipts, invoices, sales slips, gas mileage, and travel expenses.

- Maintain monthly spreadsheets reflecting income and expenses. Monthly spreadsheets make yearly filing simpler and allow you a month-to-month comparison.

- Keep detailed records. Include the date of each transaction as well as details like payment method, amount, and a description of what it was for.

To be considered a professional and not just a hobbyist by the IRS (or international tax department), you must show that you have made a profit for three out of five years, according to the Tax Reform Act of 1986. Proving this gives you professional status; otherwise, you are required to deduct expenses according to the guidelines for hobbyists. See **www.irs.gov/ businesses/small/selfemployed/index.html** for detailed information. Proof of professionalism includes:

- Your experience and expertise
- Your training in the field
- Profit and loss statements
- Separate business account (separate from personal finances)
- Detailed, accurate, and clear financial records
- Proof of time at work and in your profession
- If the work is a primary or secondary source of income

Deductible expenses include 30 percent of all professional seminars and courses, office equipment depreciation, trip expenses such as gas mileage, subscriptions to professional publications, and business-related postage. If you are a freelancer working from home, you may deduct a portion of your utilities, mortgage or rent, repair costs, property taxes, and more. For more information, visit the IRS Web site at **www.irs.gov**. The Web site includes publications that provide more information on such topics as allowable expenses and the Tax Reform Act. Also see "Business Use of Your Home," IRS Publication 587, and Publication 533, "Self-Employment Tax."

If you do not have social security tax withheld from your income or paychecks and you make more than $400 (net) per year from self-employment income, you are liable for social security tax. The difference between your allowable business deductions and your income is your net income. See IRS Schedule SE, "Computation of Social Security Self-Employment Tax." If you need to hire an accountant, the expense for doing so is deductible as a business expense.

In Amazon's Digital Publishing Agreement, Section 7 describes the tax aspect of your eBook. This portion of the agreement states that Amazon is responsible for paying taxes concerning their sale of digital books. However, you are responsible for paying taxes on income resulting from the sale of digital books. Amazon will not provide a W2 for your eBook income.

What is being sold

According to the World Intellectual Property Organization (WIPO), located on the Web at **www.wipo.int**, "Intellectual property (IP) refers to creations of the mind: inventions, literary and artistic works, and symbols, names, images, and designs used in commerce." What you are selling when you make an edition of your book for sale on the Kindle is not just a book, newspaper, magazine, or article — it is a piece of intellectual property. As such, copyright laws protect it. For a comprehensive explanation of your rights, visit the WIPO page titled "Understanding Copyright and Related Rights," located at **www.wipo.int/freepublications/en/int-property/909/wipo_pub_909.html**.

Protecting the Product: Basic Copyright Law

Know that when you are publishing your work — even in electronic form — it is protected under copyright law. Copyrighting a work means you are legally protecting it by stating who owns the intellectual property. Copyright seems to be an issue that is always in the news, in some form. Educate yourself about copyright law, because ignorance of it is no excuse in court. The U.S. government Web site provides an overview of the law at **www.copyright.gov**. Under "About Copyright," click "Copyright Basics." Copyright law is equally important for publishers of print books as it is for people who are ePublishing; it is worth learning the basics before pursuing publication in any form.

Know that plagiarism involves copying words written by someone else. Unfortunately, plagiarism is easy to do and is rampant online, which is precisely why you should take extra precautions to avoid it in your own ePublications (as well as your non-electronic work, of course). It is acceptable to use ideas and facts, as they cannot be copyrighted (and should not be), but adhere to the general guideline to never repeat more than three words in a row as another person has written them. If you would like to use the words or images of another person, obtain their permission in writing. Quoting a small portion of a passage may be all right in research papers where the source is attributed, but the rights differ, even with attribution, in books and other printed materials. Current law holds that a work is under copyright until 70 years after the death of the author. If there are multiple authors, the work falls out of copyright 70 years after the death of the last remaining author. Most military and government works are public domain.

Some documents are not copyrighted, however. To check this information, you can contact the U.S. Copyright Office. If you provide as many details as possible about the information you are seeking, the office will search for the information. Copyright law changed in 1978, and some works printed before that year may fall outside of current copyright law. For more information on copyright law past and present, visit **www.copyright.gov/forms**. Even if a work is wide open for you to legally copy, use good judgment. Strong ethics and good value judgments can go a long way in the world of publishing.

A copyright notice appears on the book's copyright page. Look at a few books for examples. A copyright appears with the copyright symbol, the year, and the author's name. The copyright symbol, rather than the word "copyright," is necessary for international recognition, and protects your work worldwide. The copyright is attributed to the owner of the work, which can be the author, publishing company, or the creator or purchaser of the work. Before printing, check proof copies to ensure all numbers are in place on the copyright page; it is a good idea to check copyrights before distribution as well. *Chapter 4 covers how to set up your copyright page in greater detail.*

Since copyright automatically rests with the author or creator of a work, no registration with the Copyright Office is technically necessary. However, the advantages to registering within three months of the book's printing lie primarily in the fact that, if you are involved in a lawsuit, you will be awarded the full amount of attorney's fees and damages. If you register more than three months after publication, the reimbursement of damages and fees is only partial. A book is registered by mailing an application form (Form TX) along with two copies of the book and a fee of $30 to the Copyright Office. Visit the Copyright Office Web site to obtain the form: **www.copyright.gov/forms**. Once your application is processed, the Copyright Office sends a copy stamped with a seal and signed by the Registrar. This document contains a date and registration number. Keep this for your records. Be aware that the processing may take up to nine months or longer, as the number of annual applications processed by the office is around 600,000.

Chapter Conclusion

Getting paid is important, but so is keeping good track of your earnings. Even if you are not a full-time freelance writer or a full-time self-employed individual, keeping track of all your earnings and reporting them responsibly and punctually will save you headaches later. Record your income as specifically as you can, keep track of monthly earnings, and bring them all to an accountant when the time is right, and she will help you file appropriately.

Always be mindful of copyright law. It pays to spend at least a little bit of time visiting the Copyright Office or WIPO Web sites. Visit them prepared with several questions about copyright. As you hunt for the answers to your questions on the sites, be patient, and be open to learning new information. You may be surprised at what you discover.

Perhaps the underlying idea is to always be honest and responsible in what you do — whether it has to do with finances or fair use — and you will save yourself the risk of being penalized later.

Protecting Written Work

Educating yourself as much as possible about copyright law is vital. Unscrupulous people exist everywhere, but there are ways of protecting yourself. A unique work that you created is your copyrighted work — the act of creating it automatically gives you the rights to the work. Unless you choose to sign over your rights in a contract or other written agreement, the work remains yours. *For more information and links to resources, see the section titled "Protecting the Product: Basic Copyright Law" in Chapter 3.*

Copyright page and information

Include a copyright notice in the book's front matter. The front matter is the material that goes in the beginning of the book and precedes the story, consisting of the copyright notice, dedication, and title page. Check the front matter in several printed books to get an idea of how these are usually laid out. A typical copyright page will contain the word "Copyright," the copyright symbol © (the letter "c" with a circle around it), and the copyright

CHAPTER 4:
Creating and Publishing a Kindle eBook Using Digital Text Platform

holder's name, which is usually the author or publisher. Below that falls the following information:

- Trademark information
- Printer information (if applicable)
- Publisher membership information
- The statement, "All rights reserved"
- Country where the book was printed
- Printer address
- CIP data
- ISBNs
- Cover and book design information and attributions
- Edition number and dates
- Any other relevant publication, book, and copyright data

Unless your material is considered public domain, include the following paragraph:

> *All rights reserved. This book may not be reproduced in any form, in whole or in part (beyond that copying permitted by U.S. Copyright Law, Section 107, "fair use" in teaching or research, Section 108, certain library copying, or in published media by reviewers in limited excerpts), without written permission from the publisher.*
>
> [Note: that's you]

Before uploading content, you must make sure that you or the company or organization you are publishing under owns the rights to all material included in the book. If you are publishing a book or other work with contributors other than yourself, you will have to negotiate with the contributors to reach an agreement regarding profits from electronic sales and permissions. If you do not, you are putting yourself at risk for copyright infringement and possible lawsuits. Be sure to publish all applicable permissions on the copyright page.

Amazon's Security Measure

According to the "Digital Publication Distribution Agreement," a document that is accessible by clicking "Terms and Conditions" at the bottom of the main DTP page, the copyright holder — which is anyone who has the right to use and distribute a work, usually the author or publisher — retains all rights and ownership to work published for sale on Amazon, while Amazon retains all ownership rights to its Properties, including their programs including the DTP.

Amazon allows content creators' use of digital rights management (DRM), which are restrictions on how many times a title you have purchased can be downloaded, and by translation, how many times the copy can be shared from device to device (such as if you upgrade your iPhone and have to re-download a copy of an eBook on the new device). The option to enable or not enable DRM is available in the "My Shelf" section of the DTP. Many consumers are ill at ease with DRM because it is so restrictive. The intention of DRM is to limit piracy — the illegal sharing, printing, and copying of eBooks. Adobe℠ Reader and Microsoft℠ Reader are two commonly used software programs that use forms of DRM. The limit on the number of times a title can be downloaded is different for each book, and is determined by Amazon in most cases. Some articles report Kindle users calling Amazon's customer service to inquire about download limits, and being informed by the representative that the publisher sets the limits; however, when I enabled DRM on my title in the DTP, I was not prompted to enter a limit. The limit is currently not disclosed before or after purchase to the user, but many users report the limit to be at about six downloads. Once the download limit has been reached, the title must be re-purchased. There is currently no way for the reader of an eBook to know up front (or at all) what the download limit is on a book prior to purchasing a title.

DRM becomes problematic, as some users have noted, when upgrading devices — some previously purchased titles do not carry over to the new device, and must be downloaded again. According to an article from June 19, 2009 at **www.gear-diary.com**, written after a perplexed Kindle user upgraded his personal iPhones and attempted to re-download all of his titles to his new devices, a call to Amazon's customer service department revealed that the publisher sets the download limits. The user had unknowingly reached his download limits on several titles, and would be forced to re-purchase them if he wanted to read them. The current alternative to DRM is non-DRM open eBooks, which leaves content wide open for copying and unauthorized distribution. Hopefully the near future holds an appropriate compromise.

According to a Jan. 15, 2010 post in the DTP forums, Kindle publishers are now provided with the option to "Enable or disable DRM" on a title-by-title basis. This selection is available in "Enter Product Details" and is located below the "Series Volume" form field.

Formats

Acceptable formats

The DTP supports the following file formats for upload to the DTP:

- Microsoft Word (.doc)
- Adobe (.pdf)
- HTML (.htm or .html)
- Plain Text (.txt)
- Zipped HTML (.zip) — this is used for HTML documents with images
- Mobi (.mobi or .prc) — Mobipocket file

Idiosyncrasies of different formats

If you are using a text (.txt) document, you will not need to save the file as HTML or PDF. The DTP processes .txt formatting and allows you to preview the document before saving. Note that Amazon recommends uploading work in a single HTML file. If you are unfamiliar with HTML, you can convert a word processing file to HTML simply by using "Save As" or "Export." If you are comfortable with HTML, you can write the HTML in a text file. If you would like to brush up on your HTML or need a reference on embedding images within the document, see **http://forums.digitaltextplatform. com/dtpforums** and click on the "Introductory HTML Formatting" link under "Formatting Guide." DTP also supports unencrypted .mobi files. *For more on uploading Mobipocket files, see Chapter 5; for a basic HTML guide, see Appendix B.*

When using Plain Text files, be aware that the Kindle is designed to automatically re-flow and re-size text. Amazon recommends using very little formatting in text files and even using as few hard returns, or hard line breaks, as possible. Formatting is anything you do to alter the text, such as bold, italics, or underlining. Users, and Amazon, caution against using PDF. While they are technically listed as a supported file format, Amazon does not guarantee their conversion quality. PDF files are essentially image files, and converting an image to text is cumbersome at best. PDF is used often for hard-copy printing. The recommended conversion test is to copy a portion of text from the PDF file and into a plain text file or Word document. If the text copies without any problems, then it will probably convert well. If you are unable to copy the text, the file will most likely not convert.

Preparation for publication

Prepare your document in a single file that contains the entire manuscript or document from start to finish, from the front cover to the back cover. Many users find that the most successful conversion format is a Microsoft Word document that has been saved as an HTML document using the simple "Save As" command. This method also ensures that any images or embedded files included in your document are fixed in their proper positions. To do this, open your Word document, click "File," go to "Save As," and choose "Web Page" if you are using Word 2007, or choose "Save as HTML" if you are using Word 2003.

If you have included any columns or tables in your document, you may want to consider formatting them as text or graphic images (if you know how to work with a graphics program or know someone who does). Simple formatting such as bold and italics should be retained without any problems in the final version.

Because the text of a Kindle title is re-flowed, be sure to remove any page numbers or references to page numbers. Footnotes are best replaced with hyperlinks that navigate users to another section of your document, such as an End-notes page.

Your users will also find it very helpful if you embed hyperlinks in your text, especially true for nonfiction works. Since the Kindle includes Basic Web functionality, including links allows users to navigate to the URLs you mention with ease. If you do so, consider including a warning to readers in the book's front matter that navigating to image-laden Web sites may slow their device's functionality. Also include in the front matter, and elsewhere throughout the document where applicable, links to your Web site, online bibliography, or blog.

It should be noted that the preview you see in the DTP after uploading material is not an accurate representation of how the document will look on the Kindle. To test the format, layout, and final presentation, you would need to publish and view the document on a Kindle.

Graphics and Images

DTP accepts the following image formats:

- Joint Photographic Experts Group or JPEG (.jpg)
- Bitmap or BMP (.bmp)
- Graphics Interchange Format or GIF (.gif)
- Portable Network Graphics or PNG (.png)

Cover images can only be TIFF (.tif) or JPEG format. If you upload a .doc document with embedded images, according to the DTP Community Support Advanced HTML Formatting Guide for Images (document 28), DTP "extracts images from the content and replaces them with HTML <img> tags." *For a basic HTML formatting guide, see Appendix B.*

Consider where rights lie when you are selecting art and illustrations for your book, and be very careful when using images found on the Internet. If you use original illustrations that you or someone else created specifically for your book, and you have the creator's full permission to use the images, you are within rights. Ensure that you credit the source properly within the book — that is, on the cover, title page, copyright page, and anywhere else in the manuscript that is appropriate, such as with a photo credit alongside the illustrations themselves. The person who creates a work automatically owns the rights to anything he or she creates.

Alternatively, you may select to hire a content provider through Elance℠ (**www.elance.com**), Guru.com℠ (**www. guru.com**), or another source. There are a bevy of royalty-free images available through various sites on the Internet, such as stock photography sites like PhotoSpin™ (**www. photospin.com**) and iStockphoto® (**www.istockphoto.com**), and sharing sites like Creative Commons℠ (to locate images on Creative Commons, visit **http://search.creativecommons. org**). Royalty-free images are sold for a flat rate and can be used after purchasing more than once by the person who buys it. The buyer is not purchasing the rights, but rather the right to use the image. Make sure to always check the legal information and the content license agreement before using a photo, no matter where you got it from.

Some helpful tips regarding transparency in images and the essential attributes of different file formats:

- **GIF**: Images support transparency; are used for animations; can handle up to 256 colors; are best for illustrated graphics, those with large blocks of color, and few colors.

- **JPEG**: Images do not support transparency; are best for photographs and images with fine detail and a lot of color; can handle up to 16 million colors; cannot be animated.

- **PNG**: Images support transparency; cannot be used for animation; PNG-8 supports up to 256 colors and are best for drawn graphics and clip art with few colors or images with large color blocks; PNG-24 supports up to 16 million colors and are best for photographs or images with a lot of color and fine detail.

Some basic guidelines for image formatting for the Kindle:

- Any images larger than 450 by 550 pixels (px) are resized by DTP.

- Image files must be 64kb or smaller, with an aspect ratio of 9 to 11, for the purposes of screen placement.

- Since the Kindle presents images in grayscale (shades of black and white) only, your images will appear best if you convert them to grayscale ahead of time. You can do this in an image editor such as Picasa™ (download it for free at **http://picasa.google.com**).

- Increase an image's sharpness when needed, but do not over-sharpen.

- If you want an image to be full-page on the Kindle, re-size it to 450px by 550px (the first number, 450 in this case, always refers to the width, and the second number to length).

- Always size down, never size up — unless you are working with vector graphics (graphics that do not lose image quality, even when sized up to a larger size, such as text in a graphics program like Adobe Photoshop™ or Adobe Fireworks™).

- Convert images with line art or text to .png format.

Book covers

A word regarding book covers — your book should have one. If you are inexperienced at any graphic editing program, try a very simple one, like Corel℠ Paint Shop Pro, which you can find in a free trial version from **www.corel. com**. You can create a surprisingly elegant cover with very few elements, using something as simple as the careful placement of text, or text overlaying a photograph you snapped and uploaded. You can also use the tools in the free Picasa software program to create a cover. Amazon provides these guidelines for images:

- The DTP automatically resizes images larger than 450 by 550 pixels.

- Images you embed, or reference, in the HTML of your DTP document should be smaller than 64 kilobytes.

- Amazon uses an aspect ratio of 9 to 11, which means images you upload will be resized to fill the screen at a ratio of 9 to 11 (you can think of this in terms of inches if it helps you to get a clearer picture of what is happening on the screen), so unless you create your image or crop it down to reflect a 9 to 11 ratio, stretching could result in the process of resizing.

- You should be aware that there is currently a known issue in the Preview that causes images to sometimes be rotated or even re-sized on the Preview screen, even if they will not appear that way on the Kindle.

QUICK TIPS ON CREATING GRAPHICS FOR COLUMNS AND TABLES IN ADOBE PHOTOSHOP OR FIREWORKS:

- Open the document containing the text you want to create a graphic for.

- Select and copy the text by highlighting it and hitting Control (or Command on a Mac) + C.

- Open Photoshop or Fireworks.

- Create a new canvas by hitting Control (or Command on a Mac) + N (or go to File > New).

- Choose the width and height. Try not to make it more than 450 pixels wide. You can adjust the length as needed by going to Image > Canvas Size. The resolution is fine at 72 px/inch (pixels per inch). White is usually the default background; remember that Kindle screens show only grayscale (black and white), so it is best to keep the white background.

- With the text tool selected, draw a text box and hit Control (or Command on a Mac) + V to paste the copied text in the box. This is your first column.

- After sizing the first text column to your liking and adjusting the font size for readability (remember that the text should be large enough to be readable on a Kindle screen), remove the text to place in the second column by highlighting it and hitting Control (or Command on a Mac) + X to cut.

- With the text tool still selected, draw your second text column (in Photoshop you may have to deselect the text box first, or click another tool to deselect and then click the text tool again) and paste the text by hitting Control (or Command on a Mac) + V.

- Adjust the font size, column lengths, and text within each column until each column is roughly the same size. Make sure the columns have equal space around them vertically and horizontally; that there is a sufficient gutter (the space between the two columns), and that they do not appear uneven.

- Save the document as a .png (Fireworks default file extension) or .psd (Photoshop default) and then export as a GIF (best for images with few and less complex colors). Keeping the .png or .psd file will ensure you have an editable file on hand in case you need to make changes. Be creative, but spend a little time working on your cover to make sure it looks as professional as possible.

If you know someone who is well versed in graphics and image editing, see if you can make use of their skills. Remember that your cover is a marketing tool, and that people do judge a book by its cover. People tend to form an opinion on the book based on the title, and the presentation, layout, and professionalism of the cover design. But if you are willing to give graphics a go and are relatively new to design concepts, here are basic dos and don'ts:

- Make use of white space. "Positive" and "negative" spaces move the eye around and add interest, as do highlights and shadows.

- Match your colors well. You do not have to refer to a color wheel, but you should be aware of whether colors clash or complement one another. (**kuler.adobe.com** is a great color palette tool).

- Keep an eye out. Look at glossy magazine ads, especially those in artist, architectural, and high fashion magazines. Pay particular attention to covers of books that have debuted in recent years.

- Be creative and, as much as possible, unique.

- "Bad" is a very subjective, qualitative value judgment that usually varies depending on who is looking (or judging), but avoid "bad" photographs — this can mean ones that are blurry, faded, have little contrast, or other missteps. Just remember that those same criteria can be used purposefully and artistically to create a "good" image. Use your judgment as to whether a photograph is of poor quality.

The list above is not intended to be exhaustive or authoritative. By all means, experiment with graphics, but be honest with yourself when judging the final product and get a second (and third) opinion.

DTP for Mac Users

Some Mac users experience difficulty with document conversion. This is primarily because Amazon's DTP is engineered with PC users in mind, and Mac text encoding is unfriendly to the DTP system, which does not accept Unicode-encoded documents. Unicode is a system of encoding characters — letters, numbers, and symbols. One way around this is to save the file in Rich Text Format (.rtf), then to open it in TextEdit (an open-source text editor and word processor), and from there, to choose Save As > HTML. Native files produced on a Mac by Pages or TextEdit will not run through the DTP. Once your file is saved as HTML, edit it if needed in an HTML editor such as Taco, which can be downloaded at **http://tacosw. com**. Then, with images embedded in your HTML document that is finished and ready to upload, zip the entire document into a single file by selecting the files, holding down the Control key, and clicking on them. Select the 'Compress' option and a .zip file will be created. You can also use a program like YemuZip, which has a zip option that is user-friendly for Microsoft Windows™. YemuZip is free and available for download at **www.yellowmug.com/yemuzip**. Your file is now ready for DTP upload. *For a basic HTML formatting guide, see Appendix B.*

Uploading a Book Using Amazon's Digital Text Platform

Now, with your book, short story, article, or other file in finished format you are ready to upload to DTP. As a final checklist, make sure all images are embedded, columns and tables converted, and the copyright information, title page, table of contents, chapters, index, acknowledgments, and anything else you find necessary to include in your book is in a single document and saved as an HTML file.

Details to enter

"My Shelf"

Navigate to the DTP homepage, and click on the "My Shelf" tab at the top. You are ready to begin uploading your document. At this stage, it cannot hurt to do one final check over your document to ensure that everything is in place.

Add new item

Click the "Add new item" button to begin the process of uploading your work.

Enter Product Details

After clicking "Add new item," multiple form fields appear under the heading "Enter Product Details." If at any time after beginning to enter information you have to leave the document, simply click the "Save entries" button at the bottom right hand corner of the Enter Product Details form and sign out. The next time you sign back in, the new title will appear as a draft (under "Status"). Simply click the plus sign next to "Enter Product Details" to expand the form and begin entering information. If you need to delete the title and start over, simply click the trash can icon to the far right, next to "Draft."

ISBN

Type the ISBN in the field only if you have this number. If you are uploading material that is available or will be available in hard-copy format and that has an ISBN, be aware that you are supposed to use a different ISBN for print and electronic versions of the same work; however, if you use the same ISBN for the hard-copy and electronic versions, you can effectively link all versions of the same document, whether it is in Kindle, paperback, or hardback version. This method enhances sales and integrates all data concerning the book, such as bibliographic information and customer reviews. Publishers of magazines and other serial publications, such as journals, should note that there is no field for an ISSN (an International Standard Serial Number is used for serial publications, such as magazines, journals, and newspapers). If you do not have an ISBN, simply leave this field blank.

Note that an International Standard Book Number (ISBN) is a measure of inventory and identification that is issued to each individual title for shipping purposes. Different editions of the same book receive different ISBNs. A single ISBN is purchased to use on all versions of that book that are printed. For example, a single ISBN-10 and a single ISBN-13 is associated with the mass market paperback version of Stephen King's *Carrie* published in 2005 by Pocket; a different ISBN-10 and ISBN-13 are assigned to the hardcover version of *Carrie* published by Doubleday. If Pocket were to release an updated version of their mass market paperback version — for example, if Stephen King approved a different set of edits on that version — it would carry an entirely different ISBN-10 and ISBN-13 than the prior version. If you take the ISBN-10 of "0385086954" and enter it into the search field on Amazon, it will turn up Doubleday's hardcover *Carrie* published in 1990. *More about ISBNs is included in Chapter 1.*

As a matter of interest, the ASIN is something you might see with increasing frequency on Amazon. ASIN stands for Amazon Standard Identification Number, and Amazon assigns it to every product sold through the Web site. If a book has a 10-digit ISBN (an ISBN-10), the same number is used by default as the ASIN. A product that is not a book and anything lacking an ISBN is assigned its own unique ASIN. The purpose of an ASIN is similar to that of an ISBN; both numbers are designated for purposes of tracking and inventory.

FACTOID:

Newspaper subscriptions are available on the Kindle. How does that work? If you visit the Amazon.com home page and click the Shop All Departments drop-down menu (the orange one on the top left hand corner of the page), and go to Kindle > Newspapers, you will be taken to the Kindle newspaper main page. Here you will notice that you can get a Kindle subscription to any newspaper with a free 14-day trial period. Subscription rates are billed monthly, and typically range from about $3.99 to $14.99. Most of the articles that are included in the daily editions of each newspaper are included in its Kindle version. For example, *The Philadelphia Inquirer's* book details page includes a notice in the product description that the Kindle edition of the newspaper "contains most articles found in the print edition, but will not include all images and tables." It also points out that features like crosswords, classifieds, and box scores are not included.

So what if you have a serial publication that you want to make available on the Kindle? Kindle users can subscribe to newspapers, blogs, magazines, and journals. Amazon started a private, self-service Kindle beta program on Dec. 7, 2009. Prior to the launch of the beta program, I sent an e-mail inquiry and received a response from an administrator in four weeks inviting me to the private beta program. I was encouraged to sign up, then send my vendor code to the administrator by e-mail, after which point my Kindle Publishing account would be enabled for newspapers and magazines. The Kindle Publishing account is separate from the regular DTP account. *See Chapter 6 for detailed instructions on publishing a magazine or newspaper for the Kindle.*

Title

Enter the full title of your work, including any subtitles. For example, in "How to Publish a Kindle Book with Amazon. com: Everything You Need to Know Explained Simply," the portion of the title after the colon is considered the subtitle, and should be included in the title field. A long title is not necessarily a hindrance, as long it is specific and informative — elements that can encourage it to appear in a search. If there is a volume number to your publication, list it here, along with the title of the magazine, journal, newspaper, or other edition. A newspaper would not include the date in the title, because that information goes in the "Pub Date" field.

Description

Provide as complete a description as possible in this field. If you are publishing a magazine or newspaper, include highlights and features from that issue, as well as the ISSN, if you have one. Refer to descriptions from large publishers' books, and to book synopses (descriptions) that appear on the back cover or dust jacket of books. The copy should be professionally written and include the highlights and most important details of the book. Be careful not to over-exaggerate — if you build up readers' hopes and your book does not deliver to those expectations, you run a strong risk of receiving negative reviews, which can hurt your sales. The description is limited to 850 characters (this includes text, punctuation, symbols, and spaces).

The following are a few examples of what to do and what not to do:

Bad Description: This book will change your life. Several experienced authors write about different ways to be successful. [How will it change your life? What is the topic? Who are the authors? Are they co-authors? What do they write about?

What is their experience? Successful in what — life, business, or money?]

Good Description: One of the most sought-after actors of the 20th century and beyond, Johnny Depp has fascinated media and fans alike since the 1980s. Known for his unconventional roles and dashing good looks, Depp has earned international acclaim through movies such as *Edward Scissorhands*, *Secret Window*, *What's Eating Gilbert Grape?*, *Donnie Brasco*, *Pirates of the Caribbean*, and many more. His charming appearance, quirky sense of humor, and infamous affairs all have kept him in the limelight. Author Johnny Mayhem's unauthorized biography presents an insider's look at the life of this famous actor, from childhood to the present day. [Good. This provides specific details about the subject of the book and the topics discussed. The copy is written in a professional, appealing way.]

Publisher

Enter the publisher's name; if you are self-publishing and this title has not appeared in print or otherwise through a publisher, enter your name in this field. It will appear more professional to enter a publisher title or your first and last name, rather than something like "Sandra's Publishing." If you have a URL set up for your writing, publishing, or another professional Web address related to the title, publisher, or author, you can enter it here.

If you want to choose an original name for your "publishing company," it is a good idea to do a fictitious name search to make sure that the name you chose is not already in use. Technically, you do not have to worry about this too much unless you are planning to use the name to form a business entity, but it cannot hurt to check in advance first to avoid confusion later. For example, Florida residents can visit the Florida Department of State Division of Corporations Web site at **http://sunbiz.org** and click the "Look up a Business Name" link in the left navigation area. Say you want to use the name Red Hat Publishing; you would look up the name to ensure it is available. If you are self-publishing, you might choose to come up with a title that sounds more professional, or simply use your last name; for example, Smith Publishing.

Language

Enter the language this document is published in. If it is a translation, enter the language it appears in here. The place to mention that it is a translation and other relevant details (about the original language and author, for instance) is in the description.

Publication date

Enter the date of the original publication. If this Kindle edition is the first time your title will be published, enter today's date in this field. If the title has appeared in another format, enter the first date it was published.

Categories

Under "Categories," click the "Add/Edit" button. This brings up a box that allows you to select the primary topics your title concerns. You are allowed a maximum of five categories. Scroll through the entire list to check all appropriate categories. Clicking on the plus sign expands the category; click on the open circle of relevant categories only to select them. A green arrow appears indicating the topic is selected. Once you have selected all appropriate categories, click the "Add Categories" button in the center of the box to send the categories to the "Selected Categories" area. Selecting multiple topics helps to add detail to your title's data, and helps it to appear accurately in a search. The "Delete" button allows you to remove unwanted or mistakenly selected categories. Once you are satisfied with your selection, click the "Confirm" button to close the window and return to "Product Details."

Author

Under "Authors," click the "Add/Edit" button. A box appears titled "Add/Edit Contributors." In the drop down menu, you can select from the following categories: "Author," "Editor," "Illustrator," "Narrator," "Photographer," "Foreword," "Introduction," "Preface," and "Translator." Type the person's name in the blank field on the left, and select the most appropriate designation from the drop down menu. A "Remove" button allows you to delete an entry. Click the "Add Another" button to create a new field; the maximum allowable number of entries is ten. Be sure to add all relevant parties to the list, including all authors, contributors, illustrators, photographer, editors, and anyone else whose category is listed here or who contributed content to the title (including someone who wrote a preface or afterword). It is important to provide attribution to as many contributors as possible. Any persons beyond the allowable ten should be specified somewhere in the book copy, along with their contribution. When you are finished, click the "Done" button.

Search keywords

Five to seven descriptive, relevant keywords is the pre-ferred target range. Keywords allow users to find your title in a search. Think of the words or phrases that best describe your content, and separate them with commas. For example, if you are writing a book on health care for children, some of the keywords (and key phrases) you might use would be *children's health, health care, pediatrics, caring for children*, and so on. Web marketing professionals base entire careers on optimizing products and Web pages for search, making products and services accessible to Internet searchers. On the Internet, there are billions of products and services to be seen and voices waiting to be heard; the trick is to be able to make your product or service as easily locatable as possible. Among those in the industry, the prevalent reasoning goes that suc-cessful keywords harness the overlap between two imaginary circles; in one are the words and phrases that are most likely to send searchers to your title, and in the other are the words and phrases most commonly used by Amazon searchers.

If this sounds baffling, try starting with a free Google key-word tool, located online at **https://adwords.google.com/select/KeywordToolExternal**. Select the radio button for "Descriptive words and phrases," and then type in a descrip-tion of your title's content to get a feel for how keywords are generated. Also, visit the Amazon help page for key phrases, located at **www.amazon.com/gp/phrase/help/help.html**. An example of a key phrase page can be found at **www.amazon.com/phrase/Johnny-Depp**. In this case, the page shows results for all products containing phrases related to Johnny Depp. Another option for finding keywords is to look at the keywords and key phrases associated with Amazon titles similar to yours.

Product image

Click the "Upload Image" button to upload your title cover. A box appears on clicking the button that allows you to upload the image from your hard drive. A notice provides a link to "Amazon's Product Image Guidelines" and warns you that images must be in either JPEG (.jpg) or TIFF (.tif/.tiff) format, and that they should be at least 500 pixels on the longest side, but that 1200px on the longest side is the preferred size. Click the "Browse" button to select the file from your computer to upload. Once you have selected the appropriate file, click the upload button. If you are not satisfied with the image you have uploaded, or have uploaded the wrong file in error, click the "Upload Image" button again to select a different file.

Edition number

Sometimes books are reprinted, sometimes with revisions. If this is the first instance of your title, type "First Edition" or simply "1" in the "Edition Number" field. Designate any subsequent editions appropriately, for example, "Third edition." If you have revised and are updating the title, it is considered a new edition. Be sure to include this information in the description as well; for example, you may want to include a sentence like, "This is the third revised, updated edition." For magazines and other serial publications, leave this field blank, unless you are publishing a special edition; for example, a magazine edition featuring photography highlights of the year might write, "2009 Special Assignments Photography Edition." Also include the edition information with the title's copyright information — on the copyright page if your title is a book. For example, a recent edition of the *Associated Press Stylebook* includes this information at the bottom of the copyright page: "First edition, August 1977" and on a new line directly beneath that, "42nd Edition, 2007."

Series title

If your title is one in a series, enter the title. For example, if you are uploading a publication that is a printing of the winners of a twice-yearly fiction competition, enter the title of the publication and the competition name, for example: "Best Literary Review Fiction Competition." A series title may also refer to an imprint title. An imprint is a specialized subdivision of a publishing house that focuses on a certain area and often carries a certain identity. For example, an imprint can be named after its editor (such as Katherine Tegen Books, a children's and young adult imprint at HarperCollins), and usually designates a certain thematic focus. Such an example might be Best Books Fiction Series.

Series volume

This field primarily concerns magazines, journals, or any other title that is issued in a series, such as a novel. (Stephen King's novel *The Green Mile* was originally issued in a series, which released the book in six parts.) An example might be "Vol. 3.1." Take a final glance over all the fields. If everything looks OK, click "Save Entries" before proceeding to ensure you do not lose any of your information. As an additional safety measure, save your product description and other relevant information in a separate document on your computer.

DRM

This is where the Digital Rights Management (DRM), discussed in detail earlier in this chapter under "Amazon's Security Measure," is either enabled or not enabled. The options available are, appropriately:

- Enable digital rights management
- Do not enable digital rights management

If you choose to enable digital rights management, your file will be converted after clicking "Save entries" to adjust your content appropriately. Once your document is successfully converted, you will be prompted to click the "Preview" button. The preview document will appear in a format resembling the Kindle. It is important that you page through all the screens to check all the content. If you notice any errors, note the link at the top of the preview area that reads "Download the HTML." Click this to download the file, make adjustments as needed, and re-upload. If you need to make changes, be sure to check the preview file again after uploading to ensure that everything displays properly. Some common errors that appear are extra spacing between sentences and stray punctuation.

Note that after making this change, you will need to confirm your rights to upload the content for distribution. Click "Save entries" to complete the process. Note that if you have already published your content and are changing your file, even in the process of re-converting it to enable or dis-enable DRM, you will have to re-publish your work by clicking "Publish." If you do not take this step, your title will be listed as a draft. After re-publishing, your content will not be buyable for 24 to 36 hours. Generally, the content will become available in one to two hours. The delay is caused by the time it takes Amazon to update their global catalogue, add the content to the search index, and update any other information that you may have changed prior to re-publishing, such as the product detail page.

Check Your Details and Upload

Upload and preview book

The next step is to confirm the content rights. You are prompted to choose from two options: "Worldwide rights — all territories" and "Individual territories — select territories." If your rights are limited, you should already know; if you have published the title under contract with a publisher or other company that holds foreign rights, you should have already checked with them or checked your contract to find out if you are within your rights in publishing the title to begin with (*see Chapter 2, "Company/Publisher Information"*). Depending on your contract, publishing for the Kindle may be something only your publisher is within rights to do.

Once you check the box to confirm you have the rights to upload the content and click the "Save entries" button, you may proceed to Step Three, "Upload and Preview."

Media location

Click the "Browse" button to locate the file on your hard drive that will be uploaded. Choose the final document and click the "Upload" button. The document will process, and should only take a few moments for users with fast Internet connections to upload. Once the document is processed, a notice will appear in a blue box letting you know that the conversion of your document has been successful.

Preview

Click the "Preview" button to preview your converted document. Take the time to look through your document carefully, and take note of any conversion problems or other errors. Of course, a best practice would be to compare the preview version with your original version and do, as much as possible, a side-by-side comparison. But watch out especially for such things as extra spacing and stray punctuation marks. Keep in mind that a known issue at the time of this writing is the improper display of images, which may appear differently (rotated or stretched) than they will on the actual Kindle device. The "Product Preview" screen that appears approximates a Kindle; you must click the left and right arrows at the top to navigate through the document. One problem you might encounter is an error message warning you there is a problem previewing your document, and which says to try again later; in that case, simply click the left arrow to return to the previous page and then navigate forward again.

Common errors might appear with line breaks and spacing. You have the option, via a link at the top of the "Product Preview" window, to download the HTML document your file has been converted into. Do this only if you are comfortable working with HTML. *For a basic HTML formatting guide, see Appendix B.* Otherwise, make the necessary changes in your document from your computer, save it, and then re-upload it by clicking "Browse" and navigating to the file (be sure it has already been saved and closed) and clicking the Upload button again.

If you have saved and closed the file on your computer and click "Browse," but the "Upload" button remains grayed out, you will have to "trick" the system into allowing you to upload the newly updated document by clicking "Browse" and selecting a file of a different type than the file you just uploaded. For example, if you originally uploaded a Word (.doc) document, found errors, and corrected them in the document as your previewed the first version, make sure your corrected final document on your computer is saved and closed, and then click Browse and navigate to a file with a different extension, even one that is obviously wrong, such as a .jpg or .xml file. Click "Upload," and the system will give you an error message, or, depending on the type of file you uploaded, will go ahead and open up the "Product Preview" window. You can then close out the preview window, if one comes up, click "Browse" again, and navigate to the corrected, final document. Click "Upload" again, and preview the document carefully. You can repeat this process as often as necessary until you are satisfied with your final document.

Enter price

Many considerations go into setting the price. You should enter a price you think is fair, but that will attract buyers. But you should also take into consideration the work's value and the pricing of the competition. Check out at least five or six comparable titles — they should be similar in length, form, content, and subject matter — and see how your price measures up. Do you need to adjust it? Keep in mind that many bestselling titles whose hardcover versions retail for $26.95

(or more) in the book store sell for $9.99 in their Kindle incarnations. Once you are satisfied with the price you have set, click "Save entries" to save this information and proceed to publishing your document.

On Jan. 15, 2010, Amazon administration posted new pricing guidelines in the DTP forums. The list price for a title now cannot exceed a print edition's lowest suggested retail price (set by the publisher), if one exists. Also, any digital books that are between three and 10 megabytes must be priced at $1.99 at least, and file sizes of more than 10 megabytes must be priced at $2.99 at least.

Publish

After entering the price, the "Publish" button in the top right corner of the screen becomes available (in other words, it is no longer grayed-out and becomes clickable). Click "Publish," and wait as your document uploads. It can take 12 to 72 hours for your content to be live in the Kindle Store, and possibly even longer for your bibliographic data, cover art, editorial, category, and sales ranking information to arrive on the product detail page for your title. Until your title is processed, you will see a message under the "My Shelf" tab in the DTP that reads, "Publishing [Your Title]. Your book is currently under review by the Kindle Operations team…"and so on. The message warns that title updates take longer to process.

Troubleshooting, Editing, and Removing Published Content

Resolving problems

Many problems are caused by incorrect file formats, or files that have trouble processing, like PDFs. If you are on a Mac and experience trouble uploading your files, refer to the section earlier in this chapter titled "DTP for Mac Users." If you are experiencing a problem or issue not covered in this chapter so far, visit the DTP support forums to see if your topic has already been covered (it is likely that it has). View the support forums at **http://forums.digitaltextplatform.com/dtpforums/forumindex.jspa**. On the main forum page, you can choose "Browse the Knowledge Base" or "Discuss in the Forums." If no answers turn up, try performing a search through an engine like Google. There are many other forums and blogs available that address the most common problems experienced in the DTP.

Revising published work

You can download published content, if necessary, by navigating to the DTP homepage and clicking the "My Shelf" tab. Click the plus sign next to the document you need to modify, and then click the plus sign that appears by "Upload and Preview Media." Click "Preview" and "Download"; you will then be prompted to save the file to your hard drive. The file downloads as a .zip file, and is in HTML. Save the file and then navigate to it on your computer. Of course, if you already have access to the copy, you can skip the preceding steps. If you are not familiar with HTML, you will want to work from the native file anyway. *For a basic HTML formatting guide, see Appendix B.*

Navigate to your file, modify it as needed, and then re-upload it. If you downloaded the HTML file from the DTP, copy the edited file back into the .zip file you downloaded. Make sure any images and other HTML content is saved in a single folder, as there can be no content in sub-folders. Then, click the plus sign by "Upload and Preview Media," navigate to the .zip file, and click "Upload." To update content that has already been published, you must then take the additional step of clicking the "Publish" button on the DTP dashboard. After your updates are processed (this may take longer than the usual maximum of 72 hours for new content), your updates will be live in the Kindle Store.

Removing published work

To remove, or "Unpublish" content, navigate to the "My Shelf" tab in the DTP dashboard. Click the plus sign that appears next to "Enter Price" (Step Three), and uncheck the box by the bold-print words "Amazon Kindle Store." The item will be removed within 72 hours, and will disappear from the Amazon store after one week (minimum). The unpublished item will not be able to be purchased by customers within 24 to 72 hours. You can re-publish the item by checking the box by "Amazon Kindle Store" and clicking the "Publish" button. Republished items appear in the store within 12 to 72 hours.

CASE STUDY: INDIE WRITER GOES THE WAY OF THE EPUBLISHER

Web site: www.aprillhamilton.com
Blog: aprillhamilton.blogspot.com
E-mail: indieauthor@gmail.com

April L. Hamilton is an author, blogger, Technorati Blog Critic, leading advocate and speaker for the indie author movement, and founder and Editor in Chief of Publetariat, the premier online news hub and community for indie authors and small imprints. She has spoken at the O'Reilly Tools of Change conference and the Writers Digest Business of Getting Published conference, and has also judged self-published books for competitions run by Writer's Digest and the Next Generation Indie Book Awards. She has been quoted in The Wall Street Journal, MSN Money *and* The Washington Times, *profiled by ABNA Books and The Writing Cast podcast, and her book,* The IndieAuthor Guide *(to be released in an updated and revised edition from Writer's Digest Books in November 2010), has received favorable mention on* CNET *and* The Huffington Post.

Ebooks are exploding, and the Apple iPad is about the most hotly anticipated device since the iPhone, in no small part due to its eReader possibilities. EBook self-publishing is going mainstream; more and more "name" mainstream authors are electing to publish their own eBooks all the time, and many more will follow thanks to Amazon's new 70 to 75 percent royalty offer on Kindle books. The eBook format is also giving rise to increased publication of short fiction collections and poetry — formats long thought to be near extinction in the commercial marketplace. All of this amounts to a viable channel for new voices.

Maybe, if we're lucky, we're on the verge of a real renaissance in literature. EReaders aren't timely in the sense of fulfilling a pre-existing market demand, but they offer the same advantages the MP3 player did when it first came on the scene: convenience, portability, cost savings on media, and the "cool" factor. The eReader is a product that has the potential to revolutionize the way people buy and consume printed material, and the revolution is only beginning. The iPod™ is scarcely more than a decade old, yet in that time frame it has been a major factor in driving acceptance of digital music as well as rejection of physical copies of music. The same is just beginning to happen with books, periodicals, and other printed materials thanks to the Kindle, Sony Reader, and other devices.

EReaders are a godsend for avid readers, academics, and professionals in fields like law and medicine. There's no more need to carry around heavy arm loads of books, research materials or technical papers; no more worry about losing or forgetting this or that crucial bit of paper; when you can carry an entire library's worth of printed matter in a device lighter and thinner than a large-format trade paperback. While it's still possible to lose your device, it's pretty simple to get your content back with regular backups or through use of the Kindle's Whispernet. And I can tell you this much from experience with friends and family: Once someone makes the switch to an eReader, just as when they make the switch from CDs to an MP3 player, they rarely go back to their old, analog habits.

Things have changed a lot just in the past couple of years, and it seems to me that self-publishing is rapidly going mainstream. The stigma has been there for decades, yet just in the past year the level of industry acceptance of self-publishing, at least here in the U.S., has increased dramatically. We're seeing very successful, mainstream-published authors such as Stephen R. Covey (he authored *The Seven Habits of Highly Effective People* and the similar books that followed), fantasy author legend Piers Anthony, and crime novelist J.A. Konrath self-publishing and being very open about it. Konrath, who is only self-publishing Kindle editions of some of his titles so far, is a particularly surprising case because he was very vocal in his anti-self-publishing stance on his blog as recently as 2008. In January 2010, I read that Martin Amis and Ian McEwan have elected to go indie with eBook publication of their back catalog titles, rather than have them re-released through a mainstream publisher, largely because they can earn a 70-75 percent royalty going the indie route. Standard eBook royalty percentages offered by traditional publishers only run about 25 percent, or sometimes 50 percent for more established authors.

As to the "why," I'd say it's a combination of factors. First, there's the fact that trade publishing is in crisis. Between struggling brick-

and-mortar booksellers, the industry's unsustainable returns policies, the pressures brought about by the eBook revolution and the recent flood of inexpensive and free eBooks, the Google Books settlement, and their already historically narrow margins, trade publishers are facing simultaneous, unprecedented challenges. As a result, they've become more risk-averse than ever and have reduced author advances and marketing budgets, all of which lead to a much less author-friendly environment — especially for debut authors. The second major factor consists of recent improvements in eBook and Print-on-Demand technologies that have made it possible for self-publishers to duplicate the quality of mainstream-published books without having to pay large upfront fees to a vanity press or sign away their publication rights. EBook publication is even simpler, and authors who publish their own eBooks enjoy royalties 25 to 50 percent higher than those whose eBooks are published by mainstream publishers. The third major factor is Amazon, which lists self-published books right alongside mainstream-published books and offers independent authors all the same sales tools and marketing opportunities as mainstream authors. It's no surprise that Amazon is now positioning itself as a publishing channel alternative to mainstream publishers, and I don't doubt the move will be entirely successful for the company. The final factor is the decline of the chain, brick-and-mortar bookstore. Chain, brick-and-mortar booksellers have always been reluctant to stock self-published books, but now that those booksellers' market share is shrinking day by day, it's no longer critical to get one's book stocked by them. All of these factors are merging to make indie authorship a very attractive alternative to mainstream publication.

The decision of whether or not to self-publish is one that ought to be made on a manuscript-by-manuscript basis. It all comes down to three questions: First, what is the author's goal for the book? It's not always just about sales; sometimes it's about building readership or supporting a related activity, such as a speaking tour. Second, can a traditional publisher help the author reach her goal more quickly and effectively than the author could do herself? And finally, is the traditional publisher *going* to do the things that will help the author reach her goal more quickly and effectively than the author could do herself? For most authors, debut authors in particular, the answers to the latter two questions are increasingly "no."

However, a niche manuscript can be a great match for an independent, niche publisher. For example, Writer's Digest™ Books is publishing a revised and updated edition of my originally self-published book, *The Indie Author Guide*, in Fall 2010. In this case, I decided that since my goal is for the book to be made available to as many indie authors (and potential indie authors) as possible, WD Books would help me reach my goal more quickly and effectively than I could do on my own. Writer's Digest is an established and trusted brand among authors and aspiring authors, WD Books' catalog is limited to books for and about writers, WD is a forward-thinking company that is open to exploring new technologies and promotion avenues, and WD has numerous established outlets through which it promotes its authors and their books (including magazines, Web sites, conferences, classes, and book clubs).

So in this specific instance, I elected to go mainstream. However, I doubt I'd do the same with one of my novels because I think they would be lost in the crowd if I sold them to a mainstream publisher — especially a publisher that's just one arm of an entertainment mega-conglomerate, such as Viacom. Smaller, independent publishers have small catalogs, and as a result they have more riding on every single title they release. They're just as invested as the author in a given book's success. Since my fiction isn't really "niche" or genre fiction, however, it's not likely to appeal to a small publisher with a specific target demographic.

I started out going down the mainstream path, and didn't have much difficulty getting a strong agent. Like most aspiring authors, at that point I thought the hardest part of the process of getting published was behind me, but I was wrong. My novel *Adelaide Einstein* got a stack of glowing rejections from the big publishers' editors, all of whom offered some variation on, "Of course I love it, but the American book-buying public doesn't want comic fiction right now. Send me something darker." So I wrote a dark comic mystery, Snow Ball. My agent didn't like it and declined to go out with it at all. Life went on, and I forgot all about my novelist aspirations until a few years ago, when I entered *Adelaide* in the Amazon Breakthrough Novel Award contest on a whim.

After accumulating 36 positive Amazon customer reviews for *Adelaide*, I concluded New York editors don't have any idea what the American book-buying public wants and made the decision to go indie. I published both novels independently, first in Amazon Kindle editions and then in trade paperbacks via CreateSpace, Amazon's POD service. The experience was definitely a learning process for me, and since I have extensive experience with technical writing I decided to document all I'd learned in another book, *The Indie Author Guide*. I became an outspoken advocate for indie authorship, which led me to found **Publetariat.com**, an online news hub and community for indie authors and small imprints. The site was an immediate hit, so I documented my processes for design and launch of the site in another book, *From Concept to Community*.

I don't see much of a future for myself as a novelist in commercial terms, though I have a couple of fiction works-in-progress and hope to finish them eventually. I've come to the conclusion that promoting indie authorship and helping my fellow writers accomplish their goals through indie authorship is what I'm meant to be doing, and I've certainly found it very fulfilling.

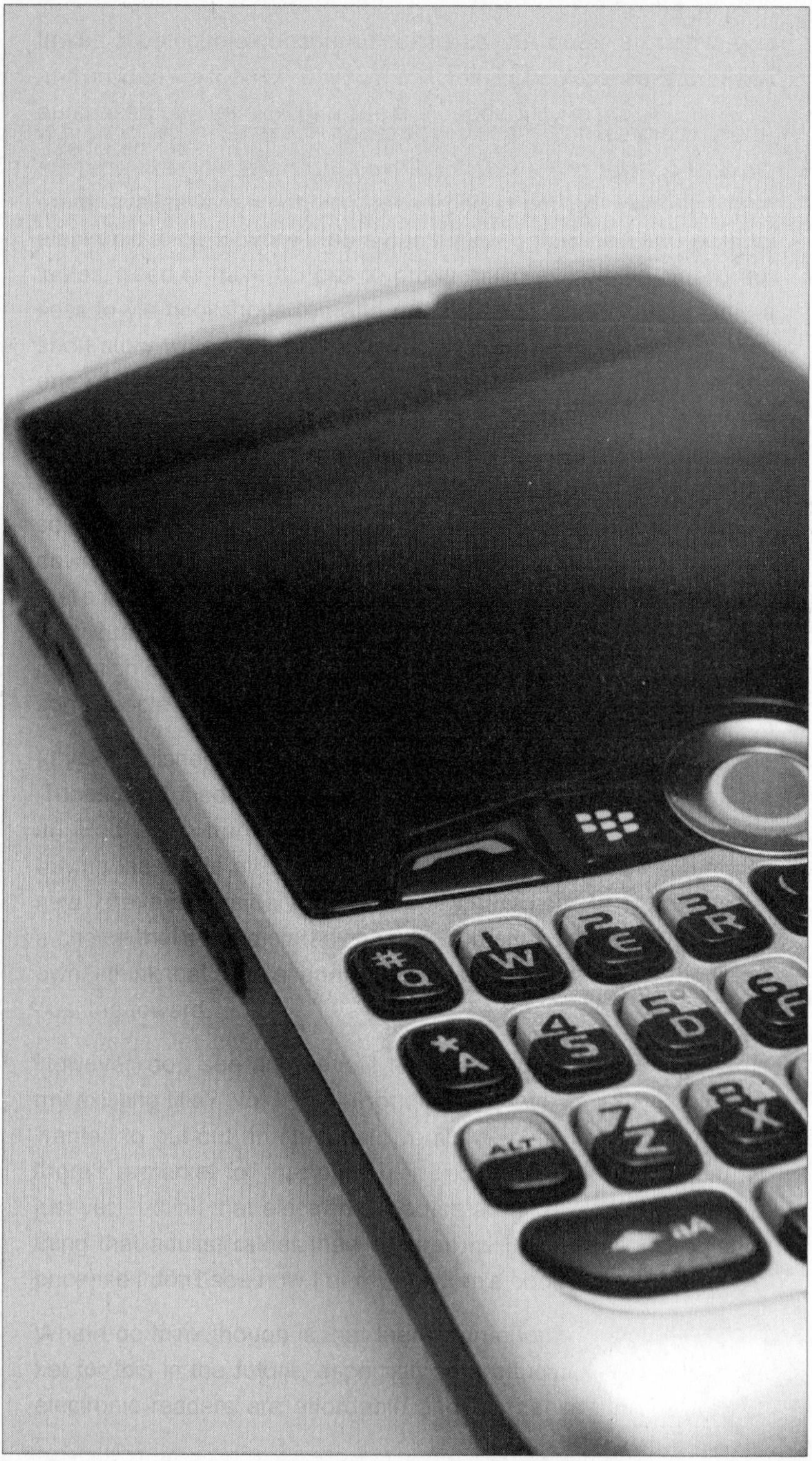

I f you will be publishing your eBook for distribution through Amazon's Kindle Store, you should be aware that your book may be downloaded and read on a device other than a Kindle. An iPhone user, for example, only needs to download an app on her phone to read her Kindle Store purchases. And now that Amazon has opened its doors to developers, newer apps like the Kindle for BlackBerry℠ open up the Kindle Store to an even wider variety of users. Read on for more details on publishing for the expanded audience of mobile phone users.

Publishing for the Mobile User

According to an Aug. 31, 2009 article in *Publishers Weekly*™, computer eBook downloads were down in July 2009 from the year's first quarter, falling from 48 percent to 40 percent. At the same time, Kindle eBook downloads went up, and downloads to the iPhone spiked, thanks to the new eBook applications ("apps") included with the 3G iPhone's release. The numbers shed light on the trends: In the first quarter of 2009, Kindle downloads were at 23.5 percent, and escalated to 28 percent in July of the same year. Sony Reader downloads went up 2 percent

CHAPTER 5:
Publishing for Mobipocket and Mobile Devices

from the initial 4 percent at first quarter. EBook downloads to other devices was up 2.5 percent from the first quarter rate of 17 percent. A Dec. 17, 2009 article in *The Huffington Post* reported that the most significant gains in book sales for 2009, a year that saw a 4.1 percent increase in book sales overall, were in eBook sales. According to the article, eBook sales had increased more than 180 percent from 2008 and, whereas they accounted for more than 1 percent of total trade book sales, they now account for 3 percent.

People now have options when it comes to reading eBooks, and they are clearly finding their preferred means of reading them in what is a nascent stage. EBooks are a relatively new concept, and the launch of eReaders and eBook apps for mobile devices means that options are expanding in an increasingly global and mobile society. So how does a writer or publisher make his or her work available for the many devices that are out there, and therefore maximize sales?

Mobipocket is both a distributor and an eBook retailer that has been owned by Amazon.com since April 2005. As is mentioned later in this chapter, Mobipocket is a platform much like Amazon's DTP, but allows users to publish content for mobile applications. Users can also send their content to the Kindle Store and mobile applications simultaneously (read on for a detailed guide on how to do this). Use Mobipocket in place of the Amazon DTP, and your work will be distributed through both Amazon's Kindle Store and to mobile and handheld devices. Your work is made available for purchase on mobile devices, like iPhones, which use apps that allow users to purchase and download eBooks onto their devices.

Located at **www.mobipocket.com**, Mobipocket files (.prc and .mobi) are supported by Amazon's DTP as long as they are unencrypted — that is, as long as the code has not been altered so as to make it unintelligible. Amazon provides a link to the Mobipocket Creator through the DTP forums. To begin, visit **http://mobipocket.com/en/DownloadSoft/ProductDetailsCreator.asp**, and download the free Mobipocket Creator.

Advantages of Mobipocket

There are several advantages to using Mobipocket. Being a retailer for eBooks, Mobipocket sends books to the Kindle Store and other eBook retailers, such as **www.booksonboard.com**. Mobipocket's wholesale distribution service is known as eBookbase. EBookbase is targeted to users of mobile devices like smartphones, PDAs, tablet PCs, and other mobile devices. Mobipocket content is encrypted with DRM, however, Kindle's DTP only accepts unencrypted files, and Amazon has its own DRM. *Digital Rights Management is discussed in further detail in Chapter 4 under "Amazon's Security Measure."*

The Mobipocket Creator allows you to create a file with HTML and image files, and has a Table of Contents Wizard, which automatically creates links to your book's chapters, using a heading extraction technology, similar to the Table of Contents generator in Microsoft Word. Converting the original file tends to be more trouble-free than it is with Amazon's DTP. The Mobipocket Creator provides users with the choice to publish the created file directly to the Kindle Store by uploading to the DTP, or to send the file through the eBookbase for distribution to multiple channels, including Amazon. The Mobipocket software tends to be stable and easy to use, with a strong support base from administration in the Mobipocket support forums. Visit the support forums for Mobipocket at **www.mobipocket.com/forum/index.php**. You will find a wide variety of topics that can come in handy when you have a question or run into a snag.

Disadvantages of Mobipocket

Mobipocket software is for Windows users only. The Creator software is incompatible with Mac computers or anyone not operating in a Windows environment. Anyone using a different operating system, like Mac, will have to download a Windows emulator to use Creator. And, whereas you can choose multiple categories for your title in Amazon's DTP, you can choose only one top-level category through Mobipocket. For example, you can classify your title only as "Computers," rather than "Computers > Electronic Commerce and Computers > Software."

It takes a couple weeks for titles to appear in the Kindle Store, and changes and updates to the book and the book details can take several days. And unlike Amazon, which pays royalties once they have reached $10, Mobipocket's payouts arrive when they reach the $150 mark (like Amazon, Mobipocket pays 35 percent royalties).

Putting the Mobipocket Creator to work

For a step-by-step guide, visit **http://mobipocket.com** and scroll down until you see the blue "Publishers" button on the left; click "Welcome Page" to be take to the "eBookBase for Publishers" page. From there, click the "Conversion Services" link under Publishers on the left to be taken to the welcome page for Mobipocket eBookBase 1.0. Under the "Start converting your content" header, click the "Mobipocket Creator Publisher Edition" link to be taken to the download page for the latest version of the Creator software. To access the tutorial, click the "Create an ebook" link in the central navigation area (or visit **http://mobipocket.com/en/DownloadSoft/ tutorial.asp?Language=EN** for a step-by-step tutorial).

After downloading the Mobipocket Creator, open the program and prepare to upload your finished manuscript. On the dashboard of the Creator (version 4.2), you should see a column on the left titled "Create New Publication," and on the right are the options to import from an existing file and to open a recent publication.

Please note that if you visit the Mobipocket home page at **www.mobipocket.com** and scroll down to the blue "Publishers" button, located in the left navigation panel, and click on "Welcome Page," you will be taken to a page titled "eBookBase for Publishers" that provides a notice stating that, effective September 2009, Mobipocket will no longer support new publisher accounts. The notice reads:

> *"Effective September 2009, we will no longer open new accounts for publishers to sell titles through the Kindle Store or Mobipocket.com. If you have an existing account, there will be no change and you can continue to upload and sell titles using eBookBase. New publishers with a U.S. address and bank account can sign up to sell eBooks in*

the Kindle Store via our self-service publishing channel at
http://dtp.amazon.com."

It is worth noting also that I followed the steps outlined in this chapter after September 2009, and had no trouble getting set up or publishing my content through Mobipocket. Of course, by the time you read this, the procedures may well have changed further, so as a default method and if you have trouble with the steps outlined in this chapter, visit Amazon's DTP to pursue publication through that channel.

Uploading a File

If you are using Creator for the first time and have not yet uploaded content, navigate to "Import from Existing File" and choose your file type from the following options: HTML document, MS Word document, or Text document. Click on the option that best suits your file type. If your file type is not listed here, go back to your file and save it as one of the above. If you downloaded the Publisher Edition, you have the option of uploading PDF documents. Creator automatically reflows complicated layouts, for example, those including embedded frames and multi-column displays. Embedded images are extracted from the PDF to PNG or JPEG formats, and headers and footers, including page numbers, are automatically removed.

The following procedure assumes you have begun with a Word file, but the process is similar for other file types. After selecting the file type, you are taken to a screen that allows you to choose a file from your hard drive. The one caveat is if that you must have the parent application for the file type you are uploading installed on your computer. In other words, if you are uploading a .doc Word file, you must have Word installed on your computer. Click on Browse to navigate to the appropriate file on your computer, and then choose the folder for the converted file to be installed. The default file location is pre-filled, and will create a folder called "My Publications" in your "Documents" folder. If you would like it saved to another place on your computer, click "Browse" and navigate to the appropriate folder location. When you are satisfied, click "Import" to being the uploading process.

The importing takes just a few seconds, depending on your machine and connections. Once it is complete, the title of your document shows in a new window, along with the file size. Clicking on the file expands a list of options on the left under "Available Actions." With the file deselected, you have the option to add another file, but clicking on the file name expands the options to allow you to remove the selected file. If you upload an HTML file you have the options not only to preview the file, but also to edit it with an HTML editor (the default is Notepad), and to make changes to the file properties. If you uploaded a Word document, you will see a few options under "View" in the left panel: "Publication files," "Cover Image," and "Table of Contents." The default selected "View" option is "Publication files."

Uploading a cover image

Click "Cover Image" to upload your cover graphic. Creator accepts PNG, GIF, JPEG, and BMP files. The size of your file in pixels is not quite as vitally important as it is for DTP, but a recommended file size might be 600px (pixels) by 800px (width by height). Remember that your image will appear on a mobile device whose screen size is much smaller than that of a typical computer.

If you plan to distribute or upload the book directly to the Amazon DTP, open the cover graphic in an image editor, such as Adobe Photoshop or Fireworks, and convert it to grayscale. Consider increasing the contrast: because the Kindle's screen processes 16 colors, adding contrast to the image can help it to appear with more depth on the Kindle and appear less "flat." The size should be about 450px by 500px.

If you are uploading the book to Mobipocket, whether or not you are also distributing to the Kindle Store you may choose to upload a color cover image. Unfortunately, high-quality results are not guaranteed for a color image viewed on the Kindle. If you plan to send the converted file directly through the Amazon DTP, you should optimize for the Kindle instead.

To upload your image file, click "Add a cover image" and navigate to the appropriate file on your hard drive. Interestingly, you can also drag and drop an image into the window. After selecting the file, the size appears to the right, along with the dimensions. Note that the actual size of the image appears in the window, as the program does not resize it to fit in the view, but rest assured that the entire image is available. Notice that in the top left, you have the option to "Preview with Web Browser." Click "Update" or "Cancel." Clicking "Update" sets the image as the cover graphic for the title selected in the Publication Files window. Clicking "Cancel" allows you to navigate back to the "Publication Files" area.

Table of contents

To begin the Table of Contents wizard, click the appropriate link in the left-hand panel under "View," then click "Add a Table of Contents." Do this only if your book needs a table of contents and does not already have one. First, consider a few things. If your book is fiction, it may be worth noting that most works of fiction do not have a table of contents; however, most short story and poetry collections do include one. Users searching for specific information can make use of tools on the Kindle and Mobipocket readers in lieu of using a table of contents. However, nonfiction, how-to, and other informational books will benefit from a table of contents.

If you have determined that you need to create a table of contents, after clicking "Add a Table of Contents," an informational field appears in the window titled "Table of Contents Generation rules." This is the TOC wizard. Above the multifield form appear the instructions, "Enter the filtering information to detect tags for which you want a link to be generated in the Table of Contents." Generation rules refers to your designation of HTML code elements, which are filtered into the Table of Contents. What you enter into the fields here will be a series of identifiers linking the (HTML) text in the TOC to the appropriate places in the document where the user will be taken to on clicking the link. Essentially, you are setting up the TOC links.

There are three columns, and three rows. The rows are simply the first, second, and third level, and the columns consist of the labels, "Tag name," "Attribute," and "Value." The column values can be broken down in terms of HTML (*see*

Appendix B for a brief guide to HTML, including usage and the definition of terms):

- **Tag name**: Here, enter HTML tags to tell the TOC what to look for in the document. Examples would be header tags, like <H1> and <H2> and paragraph tags such as <p>.

- **Attribute**: The attribute refers to a parameter setting that tells Creator what tag to look for, which is typically the class attribute. The class attribute is often found embedded in an HTML tag. For example, <p class="page">.

- **Value**: The value refers to the specific attribute that you want to direct the TOC to. Looking at the previous example, the value is "page" (alternatively, you could use "chapter").

The rows, First, Second, and Third Level, are comparable to nested lists. An example of a nested list would be:

+ Chapter (First Level)

+ Main Heading (Second Level)

+Subheading (Third Level)

+Main Heading (Second Level)

+Chapter (First Level)

A multi-tiered set-up like the one shown here might consist of subheadings within chapters. A nonfiction book (such as this one) has a chapter title, and within that chapter, main headings, subheadings, and sometimes, sub-subheadings. If your book is a novel, it will most likely only have chapter titles without subheadings. In that case, you would create a single-level TOC.

Please note: for the following section, you must have installed the Creator Publisher edition, and not the Home edition, as the features utilized in this section are not available in the Home edition.

To create a single-level table of contents, click on "Publication Files" to return to the main file window (the file list, if you have multiple titles). Click on the file you are working with, and then click "Edit with HTML editor," shown at the top of the screen. This invokes a Notepad document showing your uploaded file in its converted HTML version.

Scroll down in the HTML document to locate your first chapter or first main section. For example, if your book is a novel, scroll down to the first chapter. Assuming it is called "One," your HTML will designate a paragraph break and dictate formatting, such as the centering of the word, "One." It may look something like this:

```
<p align="CENTER">One
</p>
```

The Tag is <p>, but you will need to edit it to provide it with an Attribute and Value for the purposes of setting up your TOC in the Creator. Edit the code so it looks like this:

```
<p class="chapter"
align="CENTER">One</p>
```

Do this for each chapter in your book, changing the "One" as needed, of course. Then save the HTML file and navigate to the Creator window. Click on "Table of Contents" again, and then click the button to add a TOC. At this point, you can add your tag, attribute, and value in the wizard. Type "p" under Tag Name, "class" under Attribute, and "chapter" under Value (all without the quotes). Then click Update, and the TOC wizard will scan your HTML file, create the links, and finally, create another file titled TOC.htm. This file includes the table of contents HTML. Be sure to click "Preview with Web Browser," and check all the links to make sure they are all there. If any are missing or if you get a "Page Not Found" error message, go back and check the code for that chapter. Edit, save, and update as needed.

If you need to create a multi-level table of contents, look at your HTML. Ideally, your First Level tags will be <H1>, Second Level <H2>, and Third Level <H3>. In that case, enter H1, H2, and H3 under "Tag Name" for the first, second, and third levels. It is not necessary to enter anything under "Attribute" if you are only using the heading tags (<H1> and so on) in your HTML. If you are using them throughout the document, in places other than for chapter headings, add class to "Attribute" to specify the headings you want the TOC to pick up; otherwise, it will pick up the other header tags and provide links to erroneous places in the document. The same applies for Value: if your tags are being used only for chapter titles, enter nothing in this field. However, if you have entered a common tag like <p> under Tag Name, you will need to narrow the field down and create a class value. The class value could be something as simple as "head1," "head2," and "head3," for example. Be sure to apply them in the HTML document and save the changes.

Book settings

If you are using the Creator Publisher edition, the next option you will see on the left-hand panel under "View" is "Book Settings." If you intend to publish solely for the Kindle Store (as opposed to for the Kindle and mobile devices, or solely for mobile devices), you will not need to change anything in the "Book Settings." Clicking on this option opens up another multi-field form. The first part, "General Settings," includes options to change the encoding and book type. The Encoding option allows you to choose between "Western (Windows 1252)" and "International (UTF8)." Unless you understand what this means, leave it as is.

The "Book Type" option gives you a list of choices:

- Default
- eBook
- eNews
- News feed
- News magazine
- Images
- Microsoft Word Document
- Microsoft Excel sheet
- Microsoft PowerPoint presentation
- Plain text
- HTML
- Mobipocket game
- Franklin game

Again, if you are publishing to the Amazon DTP platform only, you can safely disregard making any changes in the Book Settings. The "Book type" is automatically set to

"Default." If you plan to publish in multiple formats, such as for a range of mobile devices and for the Kindle, select the most appropriate type from this list.

The next field in "Book Settings" allows you the check the box next to "This eBook is a dictionary." This is self-explanatory. The rest of the fields are grayed out unless this box is checked.

Clicking the "Show Advanced Settings" button opens two fields: "Palm Database Name" and "Ignore ASP directory in path." Because these are advanced settings, it is unlikely that the average user will need to employ these options. Unless you understand what they mean, leave them as-is (blank by default). Click the "Hide Advanced Settings" button to close these two fields.

Metadata

It is not necessary to fill out the Metadata form, which is the next option after "Book type" in the left menu area, but it can help to enhance search when users are looking up content. Metadata means literally, "data about data." It is structured data — data that describes the content, context, and structure of information and records — that helps users to locate your eBook in multiple search contexts. Many of the fields in the Metadata area are self-explanatory, as listed below:

EBook Title and Author

The eBook Title is the first field, and obviously, this should reflect the full title of your book. This field is limited to 150 characters. The "Author" field specifies a certain formatting: Last Name First, (comma) First Name. Separate multiple names with semicolons.

Publisher

The Publisher field is accompanied by the indicator, "This information is embedded in the book but ignored by retailers: your publisher login determines the publisher information for this book on the retail websites." In other words, what you enter in this field should ideally match what you will enter when you log in later as a publisher, but since it will be embedded in the book's information, it will be largely invisible — to the extent that it is embedded in the book's code.

ISBN

Enter an ISBN in the next field if you have one. If not, leave it blank. Common practice dictates that you use a separate ISBN for the print and electronic versions of a work, but some self-publishers advocate using the same ISBN to link all versions and enhance sales. However, Creator includes a note by this field that reads, "If you have both a paper book ISBN and an eBook ISBN, use the eBook ISBN." If your book is a serial publication and uses an ISSN, leave this field blank and include the ISSN in the Description field below. *For more information on obtaining an ISBN, see Chapter 4.*

Language

The next field allows you to select the language that the book is written in. As with the same field in the DTP, choose the primary language that is included in the copy of your title; if it is a translation, choose the language the work has been translated into. You will need to indicate in the "Description" field below that the book is a translation from, say, Hungarian.

Main subject

The next category allows you to choose the book's main subject. As you can choose only one subject here, look carefully through all the options to select the subject that best applies to your book's content. Mobipocket allows only one subject designation, with no subcategories. If you are publishing directly to Amazon's DTP, this field is not so important; you can choose multiple categories with subcategories in the DTP.

Description

The "Description" field is where you have the opportunity to summarize your book's content. Compare the descriptions on other books, looking at the back cover or dust jacket for details, highlights, and other information on what the book is about. What you write here depends partly on the genre of book you have written. If you are publishing a magazine or newspaper, include highlights and features from that issue, as well as the ISSN, if you have one. If you are publishing a novel, include the genre (literary, historical, fantasy, or science fiction, for example) and the basic storyline; try to tease the readers with interesting details that leave them wanting to read more. The copy should be professionally written and include the highlights and most important details of the book. Be careful not to over-exaggerate, and this is especially true for nonfiction — if you build up readers' hopes and your book does not deliver, you may well receive bad reviews, which can negatively affect your sales. The description is limited to 4,000 characters (this includes text, punctuation, symbols, and spaces).

Review

The "Review" field is unique to Creator (this field does not appear in the Amazon DTP) and is a chance for you to include comments, blurbs, and extracts from reviews of your book. Be aware that what you enter here does not transfer to the DTP. This space is limited to 2,000 characters.

Publishing Date and Adult Only

The "Publishing Date" field is a way for you to control your title's release. Setting the date for the future means that the book will not be made available on the retail sites until that date. You can leave the field empty and it will automatically appear as the date of publication — the date you upload your content through Creator. Below that is the "Adult Only" checkbox; check this if your book includes content not suitable for the general viewing audience, which, please remember, can include minors.

For the retail stores: metadata cover art

The next field is for your cover art. The image that you set here will show as the cover art on all the retail Web sites Mobipocket distributes through; the image you set earlier will appear as the cover art for your title. Remember that this is a metadata field, and you can choose another image here without affecting the image you set as the cover graphic earlier. The image here defaults as the cover art you set earlier; you can, but probably will not want to, use a different image. For consistency, it is a good idea to stick with the same image in both places. As with many of the other fields in the Metadata area, this option does not need much consideration if you are only uploading to Amazon's DTP.

Demo PRC file

The Demo PRC file field is one you can skip if uploading to DTP only. However, it can be useful if you plan to distribute through Mobipocket alone or through Mobipocket to the Kindle Store. If you choose to leave this field blank, Mobipocket automatically generates a demo of your book consisting of the first 5 percent. The demo is a sample readers can download to decide if they want to buy your book or not; it is essentially a preview. The Demo PRC file option allows you to upload a custom demo version of your book. Whatever you create should follow the same guidelines and format considerations as the eBook you created.

Suggested Retail Price and territory restriction

When you set the "Suggested Retail Price," remember that Mobipocket's royalties are the same as Amazon's — 35 percent. Because Amazon owns Mobipocket, authors are not docked twice for royalties from both companies, but still receive their 35 percent share on every sale. The last field is the "Territory Restriction" field; leaving this field empty allows for global distribution of your book. Only enter something here if you want to place a restriction on where your book will be made available. Be sure to be specific here.

Click "Update" to save your changes.

Guide properties

Like many of the Metadata settings, the "Guide Properties" is intended for Advanced Users. Also like many of the Metadata settings, the majority of the options under "Guide Properties" are not necessary if your book is strictly bound for the DTP. Basically, clicking the "New guide item" button invokes a multi-field form that allows you to set specific properties: Title, Type, Filename, and onclick. These fields are defined in a bulleted list, which you will see in the Creator window. However, unless you are familiar with these properties and comfortable with HTML, skip this section. *For a basic HTML formatting guide, see Appendix B.*

Build Publication: Generating the .PRC File

While you have uploaded your file and converted it to an HTML document through Creator, you still have not generated the final file for distribution — that is, the .prc file. To get started, click the "Build" icon at the top of the Creator dashboard. This takes you to the "Build Publication" screen. This screen provides you with compression and encryption options. The options you choose will depend on what you plan to do with the file, and how you want it distributed.

A basic guide on what options to choose:

- If you uploaded a custom Demo PRC file in the Metadata section, be sure to choose "No Compression" and "No Encryption."

- If you are distributing only on Mobipocket or to the Kindle Store through Mobipocket, choose "No Compression" and "Content Encryption with DRM."

- If you plan to take your file to the DTP and upload it there, choose "No Compression" and "No Encryption" so your file will be readable by the DTP.

Generally, unless your book is very long, choose "No Compression." You will notice under "Tips" on the left pane a message that reads: "High Compression may take a long time and is generally recommended for huge content size only like dictionaries and reference content." No specific document length or file size is provided, but based even on this rough idea, you should have an inkling whether this applies to your work.

Once you have selected your options, click the Build button. The settings take a moment to process, and you should be automatically taken to the "Build finished" screen.

If the build was successful and had no warning messages, you should see a screen that informs you that your eBook is ready and asks you what to do with it. Your options are: "Preview it with the Mobipocket Reader emulator," "Preview it with the Mobipocket Reader for PC," and "Open folder containing eBook." "Preview it with the Mobipocket Reader emulator" provides you with an idea of how the book will appear on various devices that Mobipocket distributes to. "Preview it with the Mobipocket Reader for PC" invokes a preview of the file in the Mobipocket Reader on your computer (PC). "Open folder containing eBook" simply brings up the file folder where your newly generated .prc is located. The .prc file icon is an eBook reader with a globe. If you are taking your file straight to the DTP to upload it, this is the file you will use. If you are taking this route, proceed to the Amazon DTP and upload this file there. *Refer to Chapter 4 for guidelines on using the DTP.*

If the build was successful, but came with one or more warning messages, the message at the top of the "Build finished" screen will read, "Build succeeded, but with warnings." To see the warning message in more detail, click the "Show build details" button. Scroll through the list of items and look for the yellow triangular warning icon. A warning can be something like "Image is too small." If your cover image is not large enough, click "Go back to the publication files" and re-upload a new cover image. Click the Build icon again, and repeat the process to generate the .prc file. Once you have taken care of all warning messages, you should be taken to the screen described in the paragraph above.

After building the file, you will be taken to a screen that provides two options: to open the file or to deploy the eBook to wholesale distribution systems. If you choose to open the file, select the appropriate radio button and click OK to bring up the file; you can then upload it to the Amazon DTP. Selecting the second option sends your title to the Mobipocket eBook-base retailer distribution platform; if you choose this option, clicking OK brings you to the "Deploy to wholesale distribution systems" screen.

Create a New Publisher Account

If you are using the Mobipocket Creator Publisher edition, you will be prompted to log in to the Mobipocket eBookbase, or to create a new publisher account. Clicking "Create a new publisher account" brings up a Web page. Before setting up your new account, you are required to have a valid e-mail address for account activation and either bank account information or the information for your PayPal account so you can begin receiving royalties.

Click "Start account creation" to begin. Fill out the fields on the following screen and click "Next" to proceed. You will be taken to the Mobipocket eBookbase Publisher Agreement. Be sure to read this carefully, as it is a legal agreement you will be entering into. Some important points of the publisher agreement are the Payment and Statement category and the Amazon Addendum category. Take note that the Payment and Statement heading specifies your payout times, which occur only four times a year — at the end of March, June, September, and December. Also remember that royalties do not pay out until they reach $150 — a far cry from Amazon's $10 minimum. The Amazon Addendum is important only if you plan to feed your eBook from Mobipocket to the Kindle Store; if you plan to take this route, be aware that you will be required to fill out tax forms, which will be sent to you from Amazon.

If you agree to the terms of the publisher agreement, click "I Agree" to proceed to the account information screen. At this screen, you are prompted to enter the royalty payment method. Choose PayPal if you expect your royalty payments to be less than $500. Create a PayPal account for free by visiting **www.PayPal.com**. Note that, while there are no monthly, setup, or maintenance fees for PayPal, you are charged a percentage on each transaction that you receive in your account. The fees vary from 1.9 percent plus $0.30 to 2.9 percent plus $0.30, depending on the amount of money being processed. Choose the wire transfer method if you expect your royalty payments to be more than $500. Keep in mind that most banks charge high fees for wire transfers, which is why you should only choose this option if you expect amounts over $500. Enter your information and click "Confirm."

The next step is to check the e-mail account for the confirmation e-mail. You will need to click the link in the e-mail to activate your account. Note that you cannot log in to the Mobipocket Creator until your account is activated. After activating, return to Creator and enter your login information. Click the Deploy Now button to send your book through Mobipocket distribution. You should receive a message informing you that the deploy was successful.

Activate Your eBook

You have successfully "deployed" (submitted) your eBook, but you are still not quite done. To activate your book, or make it available for purchase, you must set up your account information. Click on the link in Creator that appears after deploying to be taken to the Mobipocket eBookbase login page. After choosing a username and password, you arrive at the main account page. At this page, you can perform basic tasks like editing your account, viewing your agreement, and viewing sales reports.

First, visit the Retailer's list to select the online retailers you want your book distributed through (this includes the Kindle Store). Once you click on this, you will see the long list of Mobipocket retailers the eBookbase distributes to. By default, all are checked. To deselect any retailers, click the radio button by "Allow ONLY the retailers selected from the list below to sell my eBooks" and then alter the list as necessary. If you have already uploaded your title to the Kindle Store through Amazon's DTP, and now want the title distributed for mobile devices, select this option and deselect Amazon.com from the list. Choosing option No. 2 comes with a caveat; you will need to check back from time to time to see if Mobipocket has added any new retailers to the list. If they have, be sure to check their boxes. When you are done, choose "Save and close" to be taken back to the eBookbase main page.

Next, choose "Browse all eBooks." This screen shows all the eBooks you have deployed to eBookbase. From this page, you can make changes to your metadata, price, and other attributes. Click the title to be taken to the information page. Near the top of the page, there is a check box next to "Activate for Retailers." Check this box to make your book available. Unless you need to make any changes, click "Save and close" or the "Save and close this eBook information" button at the bottom of the page. Now when you return to the eBookbase main page, notice that your title shows as available ("You have 1 active eBook"). If you choose "Browse all eBooks" again, notice that you now have the option, via a link at the top of the page, to preview the book on the Mobipocket Web site or to create a promotion for the book. You may also choose "Create a Promotion" under "Promotions" on the main eBookbase page.

CASE STUDY: COZY MYSTERY WRITER SUCCESSFUL IN KINDLE PUBLISHING

Web site: www.gayletrent.com

Gayle Trent is a cozy mystery author living in Southwest Virginia. She writes two cozy mystery series, the Daphne Martin Cake Decorating mystery series and the Marcy Singer Embroidery mystery series.

I have always loved writing. My first book was published in 1999 after much trial and error. I love reading mysteries of all sorts, but the cozy mystery is my favorite genre to write because it is just so much fun. The characters are quirky, and they get involved in some crazy situations.

I have written eBooks in the past, but my book that has been most successful in ePublishing is *Murder Takes the Cake*. It has done very well on the Kindle.

Murder Takes the Cake's publisher, Bell Bridge Books, participated in a Kindle promotion wherein publishers gave free Kindle downloads of their books for a limited time. *Murder Takes the Cake* was very successful; in fact, it became a Kindle best seller. I was skeptical of the book's success, since it was free. Who would not want a free book? However, I saw the trade paperback version of *Murder Take the Cake* rise to an Amazon Sales Rank of 5,814 in books and No. 31 in Mystery Series on Jan. 2, 2010. *Dead Pan*, the sequel to *Murder Takes the Cake*, is also doing well in the Kindle Store — although not free.

As for self-publishing, investigate all your options; and remember that marketing is key. If you are not willing to market and publicize your book, and if you are not sure who your audience is, then do not self-publish.

How to Publish a Kindle Book
with Amazon.com

Introduction

On Dec. 7, 2009, Amazon launched a beta publishing program for blogs, newspapers, and magazines. To sign up, simply navigate to the Kindle magazines or newspapers section of the Kindle Store and scroll down. In the left navigation area, you will see a box that reads "Kindle Publishing for Blogs Beta: Self-publish your blog or news feed for the Kindle Store by visiting Kindle Publishing for Blogs." Clicking the link takes you to the Kindle Publishing for Blogs main page, where you can create a new account or sign in. Note that, if you already have an Amazon DTP account, your e-mail and password will not work for this area. You can use the same e-mail address but will need to create a new password.

If you are publishing a blog, you can get started as soon as your account is active. But if you are publishing a newspaper or magazine, you will need to wait for an administrator to set your account to publish for newspapers and magazines. Blogs require an RSS feed — a method of link distribution for Web site content, which can easily be added to a blog through the Gadgets section under Set-

CHAPTER 6:
Kindle Publishing for Blogs,
Newspapers, & Magazines

tings > Layout, if you are using Blogger — but the method of setting up your newspaper or magazine for Kindle publication is quite a different undertaking.

I will make of myself an example, or to be more precise, use my experience as a cautionary tale.

A week before the launch of the Kindle Publishing for Blogs beta program, I e-mailed Amazon to inquire about publishing my literary journal for the Kindle. I wanted my journal to be the first one available on the Kindle. "How exciting that would be," I thought. So I waited for a reply, and four weeks later, one came. I nearly jumped up and down with excitement: It was an invitation to the "private beta program for self-service." All I had to do was sign up for an account and e-mail the responder my Vendor Code (this appears at the top of the screen after you sign in to your account), after which he would enable my account to publish for newspapers and magazines. Shortly thereafter, my Vendor Code was in his inbox. Nine days later, he sent me an e-mail very cordially apologizing for the delay (I have to admit, I was watching my inbox with anticipation), which informed me he had set up my account with newspaper/magazine options (the default account settings are for blog publishing).

I was so excited I got right to work putting together my latest issue, which was barely two and a half weeks old, into a massive document that ended up being close to 300 pages long. I started by creating the document in the precise manner in which I had prepared my journal's Fiction Open Competition publication to send through the DTP. I thought it was long and I know poetry has a bit of trouble showing up properly on the Kindle, so I just took out the poetry. I had the document whittled down to a mere 246 pages, including the table of contents, contributor notes, internal links, masthead page, graphics, cover, and all. All told, I spent about four hours on the thing, all the while completely unaware of one very important fact that would have saved me four hours had I only known, had I only logged in to my account to look over the process a bit — publishing for newspapers and magazines is done via RSS feed. That document I had spent so long putting together was completely irrelevant, except for the cover art.

Sign up for a Kindle Publishing for Blogs account, by all means. But please be aware, before you begin formatting so much as a single keystroke of material, that it is a vastly different process from publishing content to Amazon's DTP.

After signing up for an account at **https://kindlepublishing. amazon.com**, you will need to send an e-mail to **kindle-publishing-blogs@amazon.com** requesting that your account be enabled to publish for newspapers and magazines. Be sure to include your Vendor Code in the e-mail (this appears in the top right of the screen after you log in to Kindle Publishing for Blogs). It cannot hurt to provide a description of your publication in the e-mail as well. After your account is enabled, you will receive an e-mail from an account administrator notifying you that your account has been activated to publish for newspapers and magazines. The next time you log in to Kindle Publishing for Blogs, "Kindle Publishing for Newspapers and Magazines" will appear in the top left corner of the screen after you sign in to your account.

Add Newspaper/Magazine

Click the "Add Newspaper/Magazine" link to get started. You will need to enter information to set up your publication. This is similar to how you set up your DTP account. If you need to leave this area and want your information saved, click "Save as Draft." When you log in again, your title will show up as a draft. To continue editing the draft, simply click on its link.

Publication details

The publication details you enter provide information about your publication for display and categorization in the Kindle Store.

Publication Title

The title of the publication should be the title without the edition of the issue. Issues will be available in the Kindle Store as you update your feeds, so the title should be the stand-alone title of your magazine or newspaper, and not anything associated with a certain issue. For example, *Prick of the Spindle* is the title I used, and not *Prick of the Spindle, Vol. 3.4.*

Publisher Name

Enter the name of the publisher. For most magazines, the publisher name is the same as the publication title. You should only vary from this guideline if another source or company with a different title than your magazine or newspaper prints your product. You should already know this information.

For example, *The Atlantic* (magazine) is published by Atlantic Monthly, but the publisher for *Slate* is the Slate Group.

Publication Description

This is where you can include issue details. Including information about an individual issue is a good idea only if your publication is less frequent, like a monthly or quarterly. *Prick of the Spindle* is a quarterly, so I have included details about what is included in the issue that is current; of course, every time a new issue launches, it will be important to remember to go in and change those details to update for the most recent issue. If your publication occurs more frequently, like daily or weekly, then it is best to include information about the publication itself.

Be descriptive, and include things like topics your publication focuses on, a specific coverage area, how long it has been in business, the scope of the readership, or notable writers it has published. Also mention whether anything differs in the Kindle and print (or online) editions. If there are fewer pictures, or some elements like "classifieds" are not included, you should say so here. Of course, keep your verbiage positive, like "The Kindle edition of *Lovestruck* includes most articles that appear in the print edition, but not photo essays." Include your ISSN here if you have one.

Publication Type

Here, you have the option to choose between *newspaper* and *magazine*.

Product Image

This should be an image of the most recent cover, and should be a minimum of 800 by 600 pixels. Your image can be JPG, GIF, or PNG, and must be under 1 MB. To change or remove your image, simply click the corresponding links next to the Product Image area. If you do not readily have an image you would like to use for the cover art, you can take a screenshot, or a captured image of whatever displays on your computer monitor. This can be an image of your Web site; if you choose to take a screenshot from the Web site, navigate to the most recent "cover," which usually resides on the home page.

To take the screenshot, navigate to the page you would like to capture, and hit Print Screen + Alt simultaneously; on a Mac, you can do a screen capture by holding down the Command (Apple) key ⌘ + Shift + 3, then releasing them all at once and clicking on the screen, and referring to the image file that appears on your desktop for the screen capture file. (If your keyboard has an F-key lock, make sure this is turned off.) Open a graphics program, like Adobe Fireworks or Photoshop (Paint will do the trick, too — on a Windows operating system, go to the Windows Start menu and choose All Programs > Accessories > Paint). Select Edit > Paste (the keyboard shortcut is Ctrl + V or Command + V on a Mac) and resize or crop the image if necessary. To save, go to File > Save As, enter your file name, and make sure the "Save As File Type" drop down menu us set to GIF, JPG, or PNG.

Masthead

The masthead identifies your publication and is displayed as a banner. Think of *The Wall Street Journal*'s banner logo, which appears at the top of the front cover and identifies the magazine. The masthead is not to be confused with the masthead that is a list of names on staff; rather this should contain the logo and common page or site elements that are part of the publication's branding and image. Readers should readily associate whatever is on the masthead with your publication, as it functions like an identifier.

Remember that grayscale only (black and white) appears on the Kindle. If you have a color image, convert it to grayscale in an image-editing program, like Fireworks or Photoshop, or even Paint. Acceptable file formats are JPG, PNG, and GIF. The file size must be under 1 MB, and can be no larger than 430 by 50 pixels. Ideally, make the image fit this dimension for the best screen display results.

Categories

You can select up to three categories and sub-categories here. For *Prick of the Spindle,* I selected "Arts & Entertainment," then "Literature & Fiction," and left it there. Be as specific as necessary, but when the categories stop applying to the publication, you should stop there. The categories are important because they steer your publication to a certain part of the online store. Your publication will have a better chance of turning up in relevant search results and being found by people looking for items of its category if it is correctly catalogued.

Search Keywords

The keywords field is limited to 128 characters with spaces. Choose keywords and keyword phrases that are relevant to your publication in theme, topics, and genre. *Refer to the "Search keywords" section in Chapter 4 for further details on selecting keywords.*

Language

A drop-down menu allows you to select the language your publication is written in; the choices provided are English, French, Spanish, German, and Italian. Select "Continue" to save your settings and proceed to the next page.

Feeds & Preview

This section is the most time-consuming of all the steps required to publish your magazine or newspaper for the Kindle. Especially if this is your first time working with RSS feeds and HTML or XML (Extensible Markup Language), take your time and thoroughly preview all the documents.

Step 1: Review Documentation

You will choose from three selections: Feeds & Article Format Guide, View a sample NITF feed, and View a sample HTML feed. Start with the first selection, Feeds & Article Format Guide.

Feeds & Article Format Guide

No matter how long and technical this guide or any of the supplementary materials seem, it is vital to go through everything carefully to give yourself the best chance of success at publishing your newspaper or magazine. Samples are provided for download throughout the guide. The first appears in Part 1, "Getting Started," under number 1. Click the "Sample feeds" link to download. Follow the instructions part for part (these are listed under Part 1 in numerical order); note that the instructions direct you to store the sample folder, preserving the file structure, on a location on your Web server that is publicly accessible.

Look carefully at Part 3, "Choose your RSS Feed Structure" to select how you want the Kindle to interpret your organizational structure. The categories are "Hierarchical Feeds" or "Multiple Individual Feeds." Refer to the charts for a structural representation of the two feed types. Remember that feeds will have a unique URL (Uniform Resource Locator), and your publication or site structure will be a large part of informing which feed type you should use. Compare your navigational structure to the feed types. If you have no experience with creating an RSS feed, refer to an article written by Danny Sullivan at the Search Engine Watch Web site, at **http://searchenginewatch.com/2175271**.

If you are unsure of which feed format (this is different from the feed structure) you are going to use, go to Part 4, "Choose your Article Format" and read through the information to decide. Section 4.2 explains that NITF stands for News Industry Text Format, which is used primarily by newspaper publishers and various news agencies and is written in XML. This section goes on to explain that the NITF format is supported by many publishers' Content Management Systems (CMS), and that the format is beneficial to news agencies because it allows control over the definitions of typical news elements, such as the title, byline, and publication date. Table 2 under 4.2 shows the elements of an NITF article.

Section 4.3 explains that XHTML "is a more consistent form of HTML that reduces the amount of interpretation required to format an article for use on the Kindle." But a warning is included that indicates a simple XHTML document may not be enough, as it may have to be refined. According to the guide, to ensure the best success with the XHTML document, minimal formatting elements should be used; do not use headers or footers, areas for user comments, or navigational structures. Note that you can preview your feed no matter which format you use to assess its quality and appearance. Especially if you are new to using a feed and XML or XHTML, allow yourself a learning curve and realize that getting your feed just the way you want it to look may take some trial and error. Look at "Table 3: Required Elements in an XHTML Article" for reference. Make sure all of these elements are present in your XHTML document. *A sample XHTML file is included in Appendix D; follow this as closely as possible.*

Section 4.4, "Article Best Practices," outlines standards for publication of your article that will help you to format it properly. Some of these guidelines include:

- Utilize images within articles where possible. Amazon warns that images smaller than 100 by 100 px (pixels) or 125 by 80 px may be dropped. If you use too many images, some of these may also be dropped.

- Include metadata in your articles where possible. This includes bylines, summaries, titles, IDs, and publication dates. Do not link to advertisements, articles, or external content that is not available on the Kindle.

- Media players are not supported. These include audio players, video players, and Flash content.

- Keep the body of the article in simple HTML format.

Amazon may drop an article if:

- It contains too few words

- It is a duplicate of another article

- It is outdated (if your publication is a daily, anything more than a day old should be dropped)

- The XML or XHTML is not well-formed (it has errors or is "malformed")

Section 5 provides recommendations on various aspects of the publication, such as the character encoding (UTF-8 is required), publication date, how missing article elements are handled, image formatting and placement, tables (avoid using these), and HTML formatting. The remainder of the Guidelines section contains information on troubleshooting, common questions, and some useful appendices.

Especially if you are unfamiliar with XML and will be hand-coding your content, read through the guide very carefully. I recommend printing it out, and going over it several times to make sure you understand what you will be required to do. Be sure to download the sample files to a folder on your server. It is also helpful to have the sample files open — the hierarchical-title-manifest file (only if you are using the hierarchical feed structure), and sample Section and Article XML documents. These will guide your coding.

View sample feeds

After determining from the Guidelines whether you will be using NITF or HTML, download the appropriate file by clicking the link under Step 1: Review Documentation on the Feeds & Preview page. Instructions for how to download these are provided under Part 1, "Getting Started," of the Feed Guidelines. These instructions are listed numerically, and provide instructions for uploading the file to a folder on your Web server.

Step 2: Add your RSS feed(s)

After converting your articles or other content to individual XML documents, you will proceed to check the feed addresses. Your feed address will either be the hierarchical-title-manifest document (which contains links to each of the section documents; each section document in turn links to each of the article documents), or, if you are using the multiple individual feed structure, you will provide the URL for each section, and check the address for each. Note that what is called an RSS feed is actually an XML document — not to be confused with an RSS feed you may already be providing through a feed reader. This is one or more RSS feeds, or XML documents, which you will create and place in a folder on your Web server. The links you provide in Step 2 should point directly to the publicly accessible folder on your site.

The "Check Address" button validates the correctness and presence of the directory path (URL) you provide.

With your RSS feed in place, and an understanding of the Guidelines and feed structure under your belt, move on to Step 2 to validate your feed address. Note that if you create, for example, a hierarchical file structure but really require multiple individual feeds, the publishing platform will send up a warning message and allow you to change your structure; you would simply remove the link to the hierarchical-title-manifest XML file and instead provide links to each of the section documents (feeds) by specifying the URL for each one. Note that you are required to validate each feed address by clicking the "Check Address" button before proceeding to Step 3, the preview generation.

Click "Check Address" to validate the URL. Once the address is validated, the Section Name field fills in automatically; this fills in based on the information from the <title> tag you provided in the RSS feed itself.

Step 3: Preview your feed quality

After your RSS feed URL has been validated, you can preview your feed for accuracy and completeness by clicking the buttons under either "Sample Preview" or "Full Preview and Report." The sample preview is faster and provides only a general idea of your feed appearance, while the full preview and report takes longer, about ten minutes, to generate. The "report" part of the latter option provides a list of error messages, if any errors are present in your XML documents.

To troubleshoot errors, refer to the specific error messages that appear in the report by clicking on the file title highlighted in blue under the "Error Summary" section. The XML document will open in a new tab (if you have tabbed browsing set up) or a new window. Error messages usually appear in one of two ways: If you see an entire XML document displayed, scroll to the end of the document and view the message at the bottom. Or, the full-view code may not display at all, and the error message will be at the top of the document. The line of code that displays in this window is the specific line where the error is present.

If you are using Dreamweaver™, open the document and use the "Find" command (Control + F on a PC, Command + F on a Mac) to locate the text within the line of code displayed in the error message. To check for problems within Dreamweaver: With the document open, go to File > Check Page > Validate as XML. The "Results" tab displays beneath the "Properties Inspector," along with a list of errors in the document and the line number they are on; clicking the item in the "Results" tab highlights the line of code in the document. Common errors include improperly nested tags and messages such as "White space is not allowed at this location" (this usually means that there is white space before the line of code, or that there is an ampersand symbol "&" in the text, which is interpreted as part of code in an XML document; simply replace the ampersand with the word "and"). When you have adjusted the error(s) in your XML document, save (Control + S on a PC, or Command + S on a Mac), and upload the document (the PC keyboard shortcut to upload in Dreamweaver is Ctrl + Shift + U). Then refresh the error page in your browser window to display the updated results.

Occasionally — even after fixing the problem, uploading the saved document, and refreshing the page — the message does not update in the browser. If this happens, simply exit the window and open a new one by clicking the link from the preview. The new window should display the updated results. If you still receive an error message, chances are you still have not addressed the problem. Code can be tricky, and I have solved several troublesome issues by typing the error message into a search engine and looking through the results to read up on what other users have done to address the issue. Note that user forums are often very helpful for this type of troubleshooting and should not be discounted.

Refer to the "Feed Guidelines" when necessary. This process may well be one of trial and error until you have generated a feed that satisfies the requirements. Compare the examples provided in the guidelines and refer to the appendices at the end of the guidelines for reference. Note that samples are given, as is a table (Part F) that lists HTML tags and attributes that are supported. When you have addressed your errors and warning messages, click "Re-Generate Full Preview and Report." Repeat this process as many times as it takes, until you no longer have any error messages. Note that, before continuing to the next step, it is wise to click through the entire "Kindle Original Preview," located on the right side of the screen, to check for formatting and display errors as they will appear on the Kindle.

Contacts, Schedule & Pricing

Details

Contact Information

Here, you are required to type in your full name, e-mail address, and phone number. If necessary you can remove this information and type in a new contact. There are two check boxes, one for "Business Contact," and the other for "Technical Contact." Click one or both, whichever applies. You are also asked to fill in your publication location. Note that this information should reflect the location where your offices are based. The form fields are "Country," "State/Province," and "City."

Schedule

Indicate your publication's frequency. Fields here are "Publish Frequency" (options are Daily, Daily excluding weekends, Daily excluding Sundays, Daily excluding Saturdays, Weekly, Fortnightly, Monthly, Bi-monthly, and Quarterly), "Publish Time," and "Publish Exception Days." Choose the "Publish Time" carefully, as whatever date you specify is the date that content will be pulled from your files for publication.

A note next to this field reads, "your publication's on-time performance will be monitored according to the time you specify," to emphasize that Amazon does monitor your publication's performance; a note next to the "Publish Exception Days" field indicates that your on-time performance will be negatively impacted if you fail to specify such days and content is pulled but has not been updated. "Publish Exception Days" are any days that your publication does not publish, such as Christmas and New Year's Day.

Pricing

Before entering any prices, note the line of text above the two form fields that reads "We will take these prices into consideration when we review your publication." In other words, Amazon has the final say on what price your publication will actually sell for. The two fields here are "Lowest Advertised Price per Issue" and "Lowest Advertised Subscription Price per Month." Note that prices must be in U.S. dollars. If your publication does not fall neatly within this categorization — for example, if your publication is online or anything other than monthly — figure your price as closely as possible by looking at the subscription rates of comparable publications online (either on or off Amazon. com) or by doing simple calculations.

Submitting your content

If you have come this far, you must be bristling with excitement. After all that work coding your articles (unless you are a lucky soul who is using a Content Management System) and checking and re-checking the code to get it just right, and with the thought in mind that there are still only a relative handful of magazines and newspapers available for Kindle subscription, you may think you are ready to go. Not so fast.

After clicking the last "Continue" button in the previous step, you will be asked to make a final preview of all your content. Go through each page one more time, and be sure to check over everything closely. Click the "Edit" button in the top right-hand corner if you need to change anything. After you submit your content, note that you are submitting it for approval, and not publication. (This is not like the DTP.) After you submit, you should get a response within one week from an Amazon administrator — I received a response after four weeks and several well-phrased e-mails. Until that response comes in, the status of your publication will read "Pending Approval" on the Kindle Publishing for Newspapers and Magazines home page.

CASE STUDY: SUCCESSFUL WRITER TURNS EPUBLISHING PRO

Web sites: LenEdgerly.com
TheKindleChronicles.com
TheReadingEdge.com
E-mail: LenEdgerly@gmail.com

Len Edgerly, a graduate of Harvard College (1972) and the Harvard Business School (1977) has worked as a business journalist at The Providence (R.I.) Journal-Bulletin, *as editor of an energy magazine in Casper, Wyo., and as an executive at a natural gas company based in Denver. After an early retirement, he earned a master's in fine arts in poetry from Bennington College in 2003. He has also served on arts policy boards, including the Western States Arts Federation, the Denver Commission on Cultural Affairs, and the New England Foundation for the Arts. He is the author of two titles at the Kindle Store,* A Poet's Progress at Bennington – Vol. 1 *and* Cold Turkey in Paradise: Twelve Days off the Internet at Maho Bay. *His blog, Len Edgerly: Random Reflections, is also available for Kindle subscription. He and his wife divide their time between Denver and Cambridge, Mass. He has been podcasting since December 2006 at the Audio Pod Chronicles and Video Pod Chronicles. He launched* The Kindle Chronicles, *a weekly podcast about the Kindle, in July*

2009. In January 2010, he added a companion podcast, The Reading Edge, which comprises interviews with people involved with eBooks and eReaders beyond the Kindle.

Do I think that the Kindle is at the forefront of an ePublishing revolution? Yes. It is still the market leader, by far, and I believe it is still the best eReader. After a shaky start, the current Kindle is more crash-free than the Barnes & Noble nook, and the reader interface is brilliant. Little things — such as how easy it is to search words with a powerful built-in dictionary — make reading on the Kindle a delight, in my opinion. There are more copyrighted, commercial books at the Kindle Store than at the Sony store, and third-party applications such as Calibre and KindleFeeder have extended the capabilities of the Kindle. I have ordered a Sony Reader Daily Edition, to see what I think of Sony's first wireless-enabled device. I also have a nook. I will not be able to afford buying every single promising eReader that comes down the pike, but I won't be shy in purchasing the ones that seem as if they are going to matter the most.

I have always loved to read, and I have always loved gadgets. I thought I had died and gone to the Library of Congress when the NuvoMedia released the Rocket eBook in 1998. I was sure the revolution had arrived, but as it turned out it had not yet. When the Amazon Kindle arrived in November 2007, I was wary; not wanting to look like an idiot again to my friends, I waited a whole month before I ordered one. It was love at first sight; even with that clunky original Kindle whose back cover kept falling off and you had to carry a paperclip in the cover for frequent crashes and restarts. I had a series of problems with my initial Kindles, but Amazon wonderfully kept replacing them, at no shipping cost to me. I have not bought a paper book since, except for gifts and a volume of poetry by W.S. Merwin that I was pretty sure would never be available on Kindle. I bought the Kindle 2 when it came out, and the DX, but so far I do not yet have the Kindle enabled for global wireless coverage.

It will be very difficult for anyone to dislodge Amazon from its perch astride the eBook world. I have seen recent reports that the Kindle accounts for 90 percent of the eBooks now being purchased. In device sales, the Kindle appears to have a market share of more than 60 percent, followed by the Sony Reader at 30 percent and everyone else fighting for scraps.

It is possible that Amazon will bow out of the hardware business, having succeeded in setting the standard for eBooks with the Kindle format. I just do not know on that one. I hope they open up the Kindle so it can read the ascendant ePub format natively. You can read a DRM-free ePub book on a Kindle now if you convert it to .mobi format using Stanza or Calibre. But that's a pain, and it would be much better if ePub books could be read on the Kindle.

Amazon has made hardly any mistakes so far in the deployment of the Kindle. About the only thing they have done wrong is not make enough of them. They certainly seemed to get the demand and supply right in the 2009 holiday season, which they had pretty much to themselves as the nook sold out way too soon. I am long on Amazon when it comes to the Kindle's future.

I do not think the audience for eBooks is different than that for print books. In fact, the genius of the Kindle and of eReaders following the Kindle's success is that they hew so closely to the experience of reading a traditional book. The Kindle can read your e-mail, but it is a tedious process. What it is good for is a long, careful, attentive read in your favorite chair. I think there will be a difference between the market for snazzy tablet readers such as the Apple device and the market for print books. Some people, perhaps younger readers, will not mind LCD screens for reading, and they will love having a video pop up in the middle of *Great Expectations*. Not me. I spend too much time in front of a bright computer screen all day. When I retreat to my favorite leather chair with my Kindle, my mind relaxes and my eyes say, "Finally — a break." That is how I felt when it was a paper book that I brought to my chair. E-Ink has preserved that basic aesthetic pleasure, and I think it is a crucial part of the Kindle's success.

I launched The Kindle Chronicles podcast in July 2009, determined to create and upload an episode every Friday. I had seen too much "pod-fading" by promising podcasters, and I thought a regular schedule would help to attract a loyal audience. I also wanted to learn as much as I could about the Kindle, so I thought doing an interview each week would give me a chance to have conversations with intriguing players in the emerging eBook sector. I have kept to my original structure for the podcast, with minor variations. Each episode begins with introductory comments, then Kindle news, a tech tip, the interview, content suggestions, and listener comments. Shows range from a half hour to just over an hour, and I shoot for about 45 minutes per episode. Feedburner reports that I have nearly 2,000 subscribers to the feed, and they form a very engaged audience. I receive wonderful e-mails and some audio comments that provide additional content for the show each week, and I have made good friendships with listeners from around the world.

For interview guests, I look for people who are doing lively work in what I call the Kindlesphere. I've had the leading Kindle bloggers on, such as Stephen Windwalker, Andrys Basten, Ahbi, and Bufo Calvin. Ian Freed, Amazon's Kindle vice president, has been my guest several times, along with Nicholson Baker, the *New Yorker* writer who wrote an articulate hatchet job on the Kindle and explained how that happened in our interview. I love doing the podcast because it enables me to call up just about anyone I want and talk about the Kindle. I never did hear back from Oprah, but it was worth a try.

I have just this week launched a companion podcast, The Reading Edge, at **TheReadingEdge.com**. It will comprise just interviews, and it will give me a chance to extend my curiosity well beyond the Kindle. The first episode has an interview with a Sony Reader product expert whom I met at the 2010 Consumer Electronics Show in Las Vegas. I have seven more interviews from CES that I'll be posting to The Reading Edge in the coming episodes.

51.35
31.04
5.46
96.70
13.02
545.50
90.64
88.88
+0.26
+1.30
+0.58
+0.02
+2.12
+0.40
+9.50
+0.78
+0.76

The basic principles of marketing and promotion remain primarily the same for an eBook as they do for a book in print. The main difference lies in the market itself — an eBook (apart from any print or POD version it may have) is most likely to be read on a screen, whether that screen is attached to an eReader, a computer, an iPhone, or another electronic device. Therefore, the best angle to take when marketing your ePublication will be to market it online. Besides getting the good word out about your publication, there are several other methods you can use to help increase your readership:

Pricing

Amazon standards

Books on Amazon tend to be priced around $9.99 — the default price for books that are currently on the market as hardcover editions. Books that are on the market as trade paperbacks often retail from $5.99 to $7.99 as Kindle editions, and mass market paperbacks are usually priced from $2.99 to $4.99. Monthly subscriptions to newspa-

CHAPTER 7:
Marketing and Promotion

pers and magazines usually hover between $9.99 and $14.99. Single magazine issues are typically anywhere from $1.49 to $2.99. A magazine such as *The New Yorker*, whose print version is priced at $5.99 per issue, sells for $2.99 per month in its electronic incarnation. Blog subscriptions retail for 99 cents for a monthly subscription with a free 14-day trial period, and stand-alone articles and other short-form content are priced from 99 cents to $2.99.

Keep in mind that Amazon often reduces prices to encourage sales, selling titles at a discount, which is usually 20 percent off the retail price. This reduction in price does not affect the 35 percent you earn in royalties, or the total profit per unit.

Selecting a price

The lower you price your book, in theory, the better its chances of selling. There are some strategies that have proven to work, in addition to the proper marketing. Just like with a top notch Web site, if people do not know it is there, nobody will visit it no matter how helpful, creative, or professional it is. The same concept applies to your eBook. Let people know it is out there and your sales will directly benefit. *See the end of this chapter for more on promoting your work; for promotion via social media outlets and the Internet, see Chapter 8.*

If you are selling an article, short story, or other short-form work, in general, price on the lower end; from 99 cents to $2.99. If you are selling a book, you can start low, and then adjust as needed. However, pricing above $9.99 is not recommended for a typical novel-length work. Price a book available in paperback from $2.99 to $7.99. If your book is only available as an eBook, use your discretion when pricing. Some books demand a higher price; especially content that is highly technical, graphics-heavy, or for which you have to pay contributors.

Amazon Sales Rank

Amazon reflects how well a title is doing through the sales rank. Sales rank is updated hourly and is calculated based on both recent sales and historical sales data. For purposes of narrowing down an item's sales results, products are ranked according to how well they are doing in their individual niches with category sales ranks. Regular sales rank is different because it reflects how well a product sells overall. An item's actual sales are not available to the public; this information is considered private and is only available to an author or publisher. Authors publishing through the Amazon DTP can view their sales reports in their individual accounts.

How Sales Rank works

How sales generate more sales

The more sales you make in the Kindle Store, the further up the proverbial ladder your title will climb; more sales mean higher visibility and an increase in your title's appearance. Not everyone will see their title escalate into the top ten, but even if you are in the top 100 or top 500, your visibility and your sales will improve dramatically. Customer reviews tend to have a strong impact on book sales. Negative reviews can mean a drop in sales, and positive reviews can lead to word of mouth recommendations and a positive chain reaction in purchases.

As a basis for moving into paper books

If an author releases a title for the Kindle only and then later decides to offer a Print-on-Demand (POD) version of her book, the sales statistics and reviews for the eBook version carry over into the print version that is offered in the regular Amazon book store portion of the Web site. All sales of the print and electronic versions of the title feed into the same data. Since Amazon has bought CreateSpace, making it one of its subsidiaries, creating a POD version of your title is what is often referred to as "seamless."

CreateSpace, located at **www.createspace.com**, offers a variety of options when it comes to formatting a book. These can be viewed in further detail at **www.createspace.com/Products/Book**. CreateSpace is also accessible through Amazon; simply type "createspace" in the Amazon search bar and click the CreateSpace link that comes up to be taken to the Web site. The page that comes up allows visitors to register and submit content for free. When you create your POD title, CreateSpace provides a free ISBN or UPC if you do not already have one. After you receive this number, you can use it for your existing Kindle title so that the two will be linked together on Amazon.

As a reference to promote subsequent titles

Reputations can be built (or destroyed) online as well as offline. The more work you publish, and the more positive reviews you garner, the more attention your work will receive. If you go to Amazon and type an author's name in the search bar, the search turns up results from all books and materials associated with that author's name. For example, if you type in "Annie Dillard," a list of her titles appears.

If you click on one of the titles, say, *The Writing Life*, you get the book details page. The page includes customer reviews, all of which can be viewed simply by clicking on the link next to the star rating under the title at the top of the page. If you scroll down, you will notice that this author has an Amazon author page. Navigating to her author page reveals a Bibliography, a section for books referencing the author, and a customer discussion area. At the bottom of the page, you might notice the link to "Author Central"; clicking on this link takes you to a page that allows authors to sign up for Author Central, a program that is still in its beta phase. Signing up certainly has its benefits — you can add books to your bibliography, add a biography and author photo, and upload cover images. Of course, the more titles you have available to add to your author page, the more credibility you will have. Keep in mind that if you are planning long-form works, such as novels, it cannot hurt to publish excerpts, articles, or other short-form pieces from time to time to add to your publications. *See Chapter 8 for details on how to set up your author page.*

Promotions

There are numerous ways to promote your eBook. Many of them have to do with establishing a Web presence and promoting your work online, *which is covered in detail in Chapter 8.* The following are some ideas for launching your new title:

Promotion tips

Setting low introductory price

Some authors choose to price their eBooks at 99 cents to begin with, and raise the price gradually as the title gains in popularity (and by translation, sales). Others try more creative ways of pricing, such as donating a portion of the profits to charity. In general, lower prices generate more sales.

Selling low-priced excerpts

To encourage sales of longer works for the Kindle, consider excerpting stand-alone information and pricing it lower. This tidbit of information is a teaser that will encourage people to buy the work in its entirety. If yours is a nonfiction work, you can excerpt some particularly helpful or little-known information from the book. If it is a novel, memoir, or other fictional work, excerpt a high-interest portion, whether it is from an escalated action point or another highlight from the book. Whatever people read in the excerpt should make them want to read the rest of the work. Excerpts often sell in the Kindle Store from 99 cents to $2.49 and are generally article-length, from one to five pages.

Serializing your title

If you have a long-form work, consider serializing it, or breaking it up into smaller sections and selling those separately from one another in the Kindle Store. As a general reference in regards to the length of a work, novelettes are typically considered anywhere from 7,500 to 17,500 words in length; novellas are 17,500 to 40,000 words; and novels can be considered anything more than 40,000 words. Of course, your work does not have to be fiction to be serialized, but it should be long-form, or at least long enough to be broken up into smaller subsections. Consider the possibilities of serializing a memoir or a nonfiction book. Each chapter should stand on its own and should end in such a manner that it generates interest for whatever comes next. A large number of notable works have made their entrance into the world of publication in this manner, such as many of Charles Dickens' novels.

You may choose to complete the work in its entirety and then begin serializing it, or write it as you go. Once your eBook has been serialized from start to finish, you can publish it as a single volume. An obvious benefit of serialization is that it generates interest for your work; if someone reads each chapter and is interested enough to want to read more, he or she will look forward to the next installment with anticipation. This method also holds great marketing appeal; create a buzz around the serialization, and hopefully you will generate enough to bring in new readers. If your work is good enough, you can rely to some extent on word-of-mouth and social marketing promotion.

Press release

Consider going to your local media for articles or reviews; this may work in your favor, especially if you are from a small to mid-sized town. As a local resident who has published a book (yes, even an eBook), you are newsworthy and interesting. Local newspaper, radio, and television stations may be interested in an interview or article, which can mean a step toward the local promotion of your title. Just as you would target publishers before submitting your material to a traditional publisher, do the same when pitching your book to the media. Targeting a dozen stations or newspapers selectively is better than blindly sending press releases to hundreds of stations. And just as you would address a cover letter to a specific person, try to do the same with the press release or phone call. *Refer to Appendix D for a sample press release; please note that this is an alternate example from that provided below.*

A basic press release follows AP Stylebook guidelines, and includes the following elements:

- In the upper left, the words "FOR IMMEDIATE RELEASE" appear beneath the letterhead. Note that this always appears in all caps.

- Drop down one to two lines and provide your contact information. Include your name, title, phone, e-mail address, and fax number, or that of your representative, if you have one (this might be an agent or secretary).

- Drop down two lines for the headline. Note that all text in the press release should be left-justified. Headlines are always in present tense and include actionable verbs. For example, "Local Author Publishes Fourth Self-Help Book".

- Drop down one to two lines and include the dateline followed by an em-dash. The dateline includes the city and state you are issuing the release from and the date you are mailing it. For example, "New York, NY, April 5, 2011 —".

- Paragraph 1 should start with an attention-grabbing sentence and answer the questions who, what, where, when, and why. This is known as the "lead paragraph."

- Paragraphs 2 and 3 constitute the body of your release and provide other details relevant to your book.

- The "recap" appears in the lower left corner and includes your book's release date and a recap of the main points of your book, such as author name, contact name and phone number, and release date.

Social networking offline

Networking means always keeping your eyes and ears open to new possibilities. There are networking opportunities in the magazines you read, professional associations you know of, people you talk with, or new publications and projects you hear about. Make the most of what is available to you. When you are thinking of networking offline, think in terms of how you can promote your work in your community and beyond. Look at the subject matter you have written about to determine where the appeal may lie. If you have written about the history of car shows, is there a car show that frequently comes to your area? Is there a historical automobile appreciation society that may be interested in promoting your eBook in some way?

If you are new to the publishing game, consider reading industry publications related to your field of expertise or your genre (for example, if you write about parenting, consider reading magazines like *American Baby* and *Parents*) to familiarize yourself with related groups and associations — you might even consider excerpting from your work and submitting an article to them. If your article is published, you can ask the editor to include a note that the article is excerpted from your Kindle book, and include the title and a link to the book page on Amazon. Keep your antennae always on the alert for opportunities. The world of publishing is vast and multi-faceted, and it is one that does hold a lot of opportunity — sometimes in the places you least expect it.

Family and friends

Spreading the word far and wide means letting your family and friends in on news of your publication — something as seemingly simple as a well-composed e-mail can be immensely effective. Word of mouth can be a powerful thing, and if you tell as many people as you can about your Kindle book, you might be surprised at how far down the grapevine news of your work might travel.

Associates, business contacts, and professional organizations

Talk to co-workers and your professional associates when appropriate. Consider holding a party to celebrate the launch of your Kindle edition, and invite associates whose interests and experience coincides with your book's topic.

Consider attending conferences with industry organizations. If you have written a children's book, you would go to a conference held by the Society of Children's Book Writers and Illustrators. Bring a flier for your Kindle edition, or if you have your book in hard copy, bring several copies to distribute. Of course, it should go without saying that you should be discerning and appropriate in the manner in which you broach your publication in conversation. It would be poor form to thrust your publication upon anyone, and could do more to tarnish your professional reputation than anything else. Know when to act, and know how to act. If you exchange contact information, for example, with an editor or publisher, follow up in an appropriate amount of time, and do so with a professionally written letter that includes mention of your meeting at the conference or other gathering.

Attend more than just one conference. Continue to network and gather information. Get involved wherever possible. Volunteer your time. If you help to plan a conference or assist in some other way for a professional organization, your professionalism and involvement will likely not go unnoticed. No matter how shy you may be, do not be a fly on the wall. Get involved and jump in to help where needed. Talk to people and exchange business cards when the opportunity arises.

You, the professional

Becoming a member of a professional organization tells the people you work and associate with that you are an industry professional who is interested in improving his work and who takes pride in what he does. For example, the most prevalent organization a children's author or illustrator can join is the Society of Children's Book Writers and Illustrators (SCBWI). One primary reason that SCBWI is so lauded by professionals in the children's publishing industry is the resources that it offers: its newsletters, conferences, and critique groups are of immeasurable benefit to members. The SCBWI newsletter announces events and conferences and offers articles on writing and illustrating craft, news from the industry, information on publishing and publishers, and more. The organization also hosts regional meetings, critique groups, and workshops, with limited access to nonmembers.

The Authors Guild and PEN are two organizations for published writers that provide a variety of career-related resources. The Authors Guild advocates for such issues as copyright protection, legal services, and fair compensation, and is for published authors. The Authors Guild membership requirements state that book authors have to have a publication from an American publisher that is established, and who have received a royalty contract with a significant advance, in which the author is the copyright holder. Obviously, if you have published only for the Kindle, you are disqualified unless, in regards to the Kindle title, you have signed a royalty contract with an established American publisher that has provided you with a so-called "significant advance": this is highly unlikely. Also eligible are coauthors, contributors, translators, ghostwriters, and freelance writers with stipulations as to the nature and frequency of published work. PEN advocates for free expression as its primary cause, and accepts members who have published "two or more books of a literary character or one book of exceptional distinction," according to the Web site. The Graphic Artists Guild is a useful resource for artists and illustrators. *See Appendix A for links to these and other professional organizations.*

Deborah Smith is an award-winning author of 33 romance, women's fiction, and general fiction novels, including the The New York Times bestselling novel, A Place to Call Home.

BelleBooks and its POD division, Bell Bridge Books, are owned by veteran novelists Debra Dixon (former vice president of Romance Writers of America), NYT best seller Deborah Smith, Sandra Chastain, and Martha Shields Crockett. When the foursome started BelleBooks in 2000 they had published nearly 100 books with major New York houses between them. As publishers, they have turned an experiment into a full-time business: BelleBooks now has more than 40 titles in print, and is rapidly expanding in 2010 with 30 new titles, plus an audio division. Switching from an early focus on Southern regional fiction, the small press now encompasses most pop genres, including nonfiction. Contact Debra Dixon, President, at **BELLEBOOKS@belleBooks.com***, Memphis, Tennessee.*

As far as eBooks are concerned, Debra Dixon and I began studying the market for eBooks more than five years ago. My fantasy novel, *Alice at Heart*, was one of the early experiments at Warner Books' short-lived iPublish imprint, for exclusive eBook content. In the past two years, with the explosion of eBook readers, eBook sales, and Print-on-Demand (POD) printing processes, we have moved full-steam into those areas. All of our titles are in eBook editions now, and all new titles are automatically and simultaneously offered as both print and eBook formats. In addition, we are using POD to greatly increase our abilities to publish more authors, more quickly, in print editions. We are keeping close track of new technologies such as the Espresso POD printer, which will allow customers to order, print, and pick up books at bookstores and other vending sites. We are also setting up our own eBook store to sell direct to readers, featuring .mobi format so that our books are available on a wide variety of personal devices, not just eReaders [Author's Note: *For a detailed discussion on publishing in .mobi format for mobile devices and to the Kindle Store, see Chapter 5*].

EBook and POD are tools that were tailor-made for our philosophy of the "sustainable backlist." After ten years of traditional print, every book we ever published via offset print runs is still in print, as we move them to the POD model. Every book republished has been reprinted. Our focus has always been toward the long haul. We are also experimenting with offering "added value" editions in eBook, meaning that the eBook edition contains new material not seen in any print edition, POD or traditional.

EBooks also make it far more affordable (and that is an under-statement!) to offer free review copies for promotions. We believe that putting free books in the hands of readers is the best way to grow our audience. Since many of our authors are new and un-known, readers are reluctant to take a chance on them, price-wise. But after testing a new author, they *will* pay to read more books by that author.

We now routinely give away hundreds of "eReview" books via well-known reader sites such as Library Thing.

Our recent Kindle promotion, which was successful beyond all ex-pectations, increased sales on the *print* editions dramatically, and in some cases selling more than ten times their weekly average.

I have watched, along with other authors, as the Kindle platform grew from a twinkle in the inventors' eyes to a fully grown and po-tentially overwhelming industry. I mean that as both compliment and a concern. What I have seen equally as an author and pub-lisher is a tremendous threat to the book-pricing model, which is al-ready under attack from massive online discounting and used-book sales, but also the potential to reach not just younger, tech-savvy readers, but mature readers who want visual enhancements and (with controls) audio capabilities to help their aging eyesight.

I believe authors, publishers, and booksellers need to launch a strong campaign to educate readers about the costs of producing their favorite authors and genres — costs that remain whether the book is published for "real" in a bound volume, or digitally. Content has value, regardless. Free or extremely cheap eBooks will undermine the already-not-lucrative business of making even a meager living for authors and publishers. We are not greedy; we are just trying to pay the bills, here.

Yes, even the big New York publishing conglomerates operate on very slim profit margins for most books; for every Stephen King and Stephenie Meyer there are a thousand small successes that delight readers but barely cover the costs of publishing; if the reading public refuses to support a reasonable price for eBooks (and all related digital entertainment technologies in the pipeline) they will literally kill the Golden Goose.

Do I think I have a different readership for my books in print than I do for my books on the Kindle? Yes. The eBook market is still, in general (very big generalization here) focused on certain types of genre fiction, much of it appealing to younger readers and/or hardcore genre readers in romance, fantasy, erotica, sci-fi, and others. When eReaders go big-time mainstream — i.e., when your grandparents are sitting in the local coffee shop reading the daily news, sports, and the latest best seller on their eReaders — then, maybe print books will become second-class citizens. But for now and the foreseeable future, print books continue to be the Gold Standard for best-seller lists and industry success.

By 2011 we will have nearly 100 titles in circulation, all of which are available in trade paperback and multi-format eBook, with nearly a dozen selected titles in unabridged audio download editions we create in-house. Our new division, Bell Bridge Books, publishes POD editions via Lightning Source (Ingram) with selected titles designated fully returnable, on a par with our traditionally printed books.

Our successes include numerous sub-rights sales to large print publishers, foreign publishers, mass market, and book clubs. Our books have won awards. One of our non-fiction authors, Susan Nethero, (*Bra Talk*) has appeared on The Oprah Show three times.

In early 2010, a highly successful promotion put five of our eBook editions (and concurrent print editions of those books) at the top of

Amazon's Kindle best seller lists (both overall and in genre categories). EBook editions of Bell Bridge Books have also been best sellers in other formats via Fictionwise.com. Deborah Smith's 2007 women's fiction novel, *A Gentle Rain* (which is also available in a Kindle edition), spent eight weeks on the Amazon women's fiction best-seller list.

We believe the returnable books model is a broken model. It made sense only in a different era, when the accepted practice was to lay down an impressive display of each new book in order to get readers' attention. There was no other way. Today, with the Internet, social media, and online stores for all major book outlets, "books on shelves" is a quaint but outdated idea.

Like most small presses, we have struggled to succeed in the outdated system. Even our traditionally printed books are often overlooked by major chains, who want massive co-opt money to guarantee shelf placement, and indie booksellers who want publishers to give them no-risk inventory. Our POD model has run up against entrenched and adamant resistance on the part of booksellers, who seem to be hanging onto the last vestige of a failing system rather than exploring new venues.

Think of the Internet as a worldwide community where everyone comes together to share ideas and information. Besides the vast array of informational sites, the Internet's "communities" consist of Web sites that encourage social interaction and networking. Some of the most prominent are user forums, YouTube℠, and social networks like Twitter, MySpace℠, and Facebook. MySpace and Facebook were once thought of as Web sites utilized primarily by teens, but they have rapidly moved away from that image, as people of all ages continue to join the vast and ever-growing online community. Not only are these "big three" social media outlets becoming friendly to a wider variety of users, but they are also being recognized by business and Web site owners as prime grounds for marketing their businesses, products, and services.

Establishing a Presence on Amazon

Posting your title to Amazon is not enough to become successful — you can have a brilliantly written book, but if you do not let people know it is there, and let them

CHAPTER 8:
Making the Most of the Internet and Social Media

know about you as a writer, then your book might never sell so much as a single copy. Contribute your background information to let users know who you are and what you have published. This is part of building trust, establishing credentials (by answering the question, *Who are you and what gives you the authority to write what you have?*), and the beginning of making yourself known as an author. Your first step is setting up your Author Central, which was formerly known as "Amazon Connect."

Author Central

Visit **https://authorcentral.amazon.com** to get started creating your profile. Simply click the "Join Author Central" button to begin setting up your profile.

Author profile

Most likely you already have an e-mail address and password established with Amazon; you do if you have already uploaded content to Amazon's DTP. Use the same information to log in to "Author Central." Once you log in, you are taken to the "Terms & Conditions page." Be sure to review these carefully; it cannot hurt to copy and save these in a separate Word file and store it on your computer for reference. If you agree to these, click the "Accept" button. After accepting the terms and conditions, you will be asked to confirm your identity. Any existing titles you have will appear on this page along with the question, "Are you this author?" As long as the title(s) that appear are yours, click "Yes, this is me" to proceed to the account creation. A confirmation e-mail will be sent to the address you logged in with; after clicking the confirmation link, you will be able to proceed with the creation of your profile. Note that your publisher will also receive an e-mail from Amazon that they will need to verify your information. Once the publisher responds, Amazon will send you an e-mail. You can visit the Account Status page to check your account status; the link to this page is provided on the page that appears when you click the link in your confirmation e-mail.

In the meantime, you can proceed with adding a biography and photo to your author profile. As the "Join Author Central" page indicates, this information will automatically be added once your publisher has confirmed your status. Proceed by clicking the "Go to Author Central" button. From this page, you can click on the links to add a photo and biography. Note the tabs at the top of the page: "Home," "Books," "Profile," "Blog," "Videos," and "Events." Adding as much information as you have available will enhance your profile as an author. Of course, it should go without saying that everything you post will reflect your image as an author. It is always in your best interest to be as professional as possible. For example, Amazon asks that your profile picture include only you in the photograph, and not any pets. Look at prominent author photographs to get an idea of the type of image you should post. If you do not have a suitable photograph, arrange to have someone take pictures of you looking your best in good lighting. Inexpensive portrait packages are available at department stores.

Biography

Click on the biography link under the "Profile" tab in "Author Central" to upload your biography. If you do not already have an author biography or are unsure of how to write one, the best reference is other author biographies. Look at Author Central biographies such as Terese Svoboda, Mary Oliver, and Leslie What. As a general guideline, biographies should be in third person (however, Terese Svoboda's is written in first person and is done so professionally; the result is a conversational, personable tone). Include information like other books published, what inspires your writing, and awards won. Keep the biography professional, and do not include information such as how many cats you have, what your hobbies are (unless they relate to your writing), or anything else that is non-professional or off-topic.

The following is a sample biography:

> *Alan A. Example established his ability to make quantum mechanics understandable to laypersons with his first publication,* Walking the Planck: Understanding the Wave-Particle Duality, *which sold five million copies in 14 different languages. His other books continue to be bestsellers, and include* Wavefunctions and You, Toward a Unified Theory, Particles Are Massy: Science for Kids, *and* The Momentum of the Atomic Orbital. *Example has won the Nobel Prize in both science and literature, and continues to write essays and science fiction novels from his mountain home in California.*

Amazon asks for a minimum of 100 characters (this includes letters, symbols, punctuation, and spaces) for biographies. The form field does not support HTML or text formatting, like bold and italics. After typing or copying and pasting your biography in the text box, click the "Preview biography" button to see your biography as it will appear on Amazon. To make changes, click "Go back"; if you are satisfied with your biography, click "Save biography." Once your biography is saved, you have the option to edit or delete it. Keep in mind that your biography is not posted to Author Central until your publisher verification goes through.

Bibliography

Once your publisher has confirmed your account and it is activated, you can add books to your account by clicking the "Books" tab. Click on the "Add more books" button. This invokes an Amazon-wide search of any titles including your name. Once the results box appears, click on any titles that you have written. Clicking "This is my book" next to a title brings up a list of any authors associated with a title, along with a button next to each name. Click "This is me" if your name appears in the list. Alternatively, Amazon allows you to add your name to the title if it is not listed, or to report a pen name. Simply click the requisite links to send this information to Amazon. Note that if you are an editor for a work, you will not be able to add the title to your bibliography. However, when setting up my information, I e-mailed Amazon to find out if there was any way I could add a short fiction collection I had edited to my Bibliography. They responded, stating that while contributors were normally required to be listed as an author in order for a title to be added to his or her bibliography, they make exceptions for the editors of collections and anthologies. They noted that the change would take effect in one to three days, and it did.

Blog

Once your publisher has confirmed your account and it is activated, you will have the capability to blog directly from Author Central (at no additional charge) by visiting the "Blog" tab. You can add blog posts in two ways: either by clicking the "Add an RSS feed" button and adding the feed address (note that this is not the same as the blog address) or by clicking the "Create a new post" address and adding a new post directly to Amazon through the box that comes up.

To create an RSS feed using Blogger, go to your blog, make sure you are signed in, and click "Customize" at the top. Make sure the "Layout" tab is selected, and click the link to "Add a Gadget." Add the "Subscription Links" gadget to your blog, following all the instructions for set-up (they are very simple; Blogger will ask you to choose what type of feed you want). You now have RSS subscription capabilities added to your blog. To find your RSS feed address, go to your blog's home page, and under the blog subscription link, click "Posts," and make your choice from the list of how you would like to view your feed. The next page that comes up should display your feed address in the URL. Simply copy and paste this into the appropriate box at Author Central to start pulling your blog feeds into your Author Central account. A message warns you that it can take up to 24 hours for new posts to appear on your author page.

Videos

Amazon now allows authors to upload video. Click on the "Videos" tab to upload your material. You can upload video in .avi, .mov, .flv, .mpg, and .wmv file formats. Why would authors want to post videos on their author pages? If you have given lectures, readings, or gone on speaking tours, video is a great way to market your work. Or, consider uploading a video of you discussing your book (no spoilers!), your work as an author, why you write, and other features of your books or writing. Amazon allows files up to 500 MB; click on the "content guidelines" link to verify that your video meets the content requirements.

Making video is relatively simple, and most new computers now come with a camera embedded. Video-recording applications vary from computer to computer, so begin by performing a search on your computer for "camera" or "video." If you do not have a camera included in your computer, you can purchase a webcam new for as low as $7.99 from **www. buy.com** or even lower by doing a search on **www.amazon. com**. Since every camera is different, read the instructions on your device to learn how to set it up and record properly.

Amazon's video guidelines request that authors not post:

- Obscene content or anything that is distasteful (examples of distasteful content might be anything that includes foul language or portrays others in a negative light)

- Advertisements or promotions

- Materials belonging to anyone besides the author

- Personal information, such as mailing addresses and phone numbers (Amazon includes URLs in this list as well)

- Price, availability of materials, or ordering and shipping information (such as methods of purchasing directly from the author or publisher)

- Comments on information available on the Author Page and book detail pages (such as book reviews)

- Plot spoilers

- Solicitations for positive reviews and votes

Events

Once your account has been activated, you will be able to add information about your lectures, speaking engagements, book tours, bookstore appearances and signings, and other events under the "Events" tab. Amazon partners with **www. booktour.com**, which tracks author engagements; any new events you post on your Author Central page will be shared with Book Tour, who also shares your information, according to the notice on the "Events" page, "around the web to local media outlets, event listing services, and other book-friendly sites."

To post a new event, simply click the "Create new event" button. A box will pop up and prompt you to enter a description of the event, the venue, the book the event is associated with, and the date and start time. Be as detailed as possible in the description field, and enter as much information as you know about the event. If you will give a lecture on your book or related themes, enter this information here. This is also the opportunity for you to add marketing copy, as long as it is relevant to the book and event surrounding it. What you enter in the description field should entice readers of your event data to come to the event. Events are a great way to generate publicity and show that you are participating as a professional in the marketing of your work. Create events for talks, signings, lectures, readings, and any other events related to your material and writing. After filling in all the fields, be sure to click "Save event." After saving the event, links are provided that allow you to come back and edit or delete event information as necessary.

Establishing a Presence on the World Wide Web

An important part of your public face as an author is the information you provide about yourself and your eBook online. Think of your Web presence as a necessary — or at least an indispensable — promotional tool that you can use to provide details for readers and potential readers about yourself and your work, upcoming engagements and tours, professional data, and contact information.

Blog

A blog is a great way to showcase your knowledge and expertise, to promote your eBook, and to get your readers in on the conversation. The two top blog providers are Word-Press™, located online at **http://wordpress.com**, and Blog-ger℠, located at **www.blogger.com**. Both providers are free. A word of caution: you should only sign up for a blog if you are willing and able to update it regularly — at the least, once per week. Blogs exist primarily to provide regular updates and are a great tool if used correctly. You do not need coding or programming experience to run a blog.

A good example of a blog is that of writer Anne Mini, whose blog is titled "Author! Author!," and is located at **www. annemini.com**. Anne Mini, born into a family with a literary legacy, has written regularly for periodicals and published her award-winning memoir, *A Family Darkly: Love, Loss, and the Final Passions of Philip K. Dick*, in 2004. Anne Mini's blog serves as the repository for her knowledge (and it is pretty vast) of grammar rules, manuscript formatting guidelines, and many other things related to writing, editing, and the ever-changing world of publishing. She updates it regularly, and often runs series of posts related to a single writing-oriented topic.

As a writer, your focus can vary. You may choose to write about your experience with ePublishing, formatting your work for the Kindle, and marketing your eBook; the submission process; your experience with agents, editors, or publishers; your subject matter or area of expertise; the intersection between your writing life and your career, or on an endless variety of other topics. Share your experience — on a professional level, of course — with your readers and you will develop your professional face.

Web site

A primary method of advertising you should look to before considering spending money on something like magazine advertising is your Web site. A Web site should be central to your public identity as an author and the go-to resource for all things concerning your publications, interviews, professional memberships, availability for speaking, presentations, and book signings, press releases, bibliography, contact information, reviews and blurbs, and any other promotional materials concerning your publications and professional life.

A Web site is a great promotional tool. It is also a reflection of you as a professional (and your Kindle book). So what goes on an author's Web site? And most importantly, how do you get started?

- Check out those of current authors to get an idea. Refer to **www.ginahyams.com/index.html**, **www. kevinprufer.com/index.html**, **www.mdbell.com**, and **www.stephenking.com** as examples.

- You may choose to set up a Web site through a blog, such as Blogger (**www.blogger.com**) or WordPress (**http://wordpress.com**), or to purchase a domain name and secure a Web host. But you will probably only want to choose the latter option if you have experience with Web design or the patience, willingness, and desire to learn. You should also have a Web authoring program, such as Adobe Dreamweaver if you choose this option. For a beginners' tutorial, visit How-to-Build-Websites.com at **www.how-to-build-websites.com/lessonOne.php**.

- Including information such as a biographical statement, relevant bylines, author photo, publications, memberships, availability for speaking and presentations, and other relevant professional experience is requisite fare.

- Include book information, such as cover art, publisher information, ISBN, book copy (the book description), target readership, publication date, and maybe a sneak peek from the first chapter.

- Include information and links as to where and how your book can be ordered, reviews of your work, synopses of individual works, and promotional blurbs by others in the industry.

- Be sure to attribute blurbs and quotes from reviewers appropriately.

- Be careful not to mix information. If you are an accountant by day and a mystery author by night, do not mix in information about your double life as a number cruncher.

- Keep focused on your life as a professional author. Alternatively, if you are a librarian in your other life, then your experience is relevant to your calling as a mystery author, especially if your work has had a direct influence on your writing. Or, perhaps you entered your job precisely because you always wanted to be a mystery author, and now you have published your book for the Kindle. Both angles are relevant.

- Include upcoming dates and a log of past ones. These can include speaking tours, book tours, signings, lectures, and other professional engagements.

Whether you maintain a Web site yourself or have someone else do it for you, it is important to keep your content fresh and relevant. Update your site often, even if it is only to post news and happenings in your industry from around the Web, or your reaction to a book you have read recently. You may consider updating content on a regular basis, like every Friday, or more often throughout the week. The more information-based and relevant your site is to your topic (which may be electronic publishing as a whole, or picture books, or your own work only), the more sites will link to yours and the more popular you will be with search engines.

It is important to impress the search engines; especially Google, which is the largest. The more relevant your site appears to the search engines, the higher your site will rank in a search for the terms (or keywords) associated with your site. Depending on how far into Web design your interest takes you, you may come across the term SEO, which stands for search engine optimization. SEO, at its most basic level, has to do with understanding the elements the search engines place importance on in terms of how they affect your site's search ranking. If you are interested in learning more about this topic and what you can do to improve your site's search engine ranking, consider subscribing to an e-newsletter like SiteProNews (**www.sitepronews.com**) or WebProNews (**www.webpronews.com**). Both sites e-mail newsletters regularly, which contain informative, up-to-date articles on topics like SEO, Web design, and search industry trends. But the search engines are not the only ones you need to impress. Keeping content fresh and interesting will keep visitors coming back to your site.

If you maintain a site yourself, you can place HTML keywords in the header portion of each page on your Web site. There are numerous templates available online for building Web sites, or you may look into a desktop publishing program, or even a Web design program like Adobe Dreamweaver or Frontpage. If you are adventurous, willing to learn, enjoy working on the computer, and the prospect of building your own Web site seems exciting, read up on the different programs available before choosing one. Some of them, like Dreamweaver, are powerful, but learning to use them can take time and a little knowledge of HTML (hypertext markup language). *For a basic HTML formatting guide, see Appendix B.* While HTML is not required for Dreamweaver, it is helpful, as it gives you an edge, allows you to troubleshoot, and gives you greater control over your content, such as the ability to implement SEO strategies. If you are considering hiring a Web designer, start by looking at local companies and comparing them in price and quality with local freelancers, who may advertise their services in the local newspaper and online on Web sites such as **Elance.com** or even **Craigslist.org**. For an article that discusses things to consider when choosing a Web designer, see **http://articles. sitepoint.com/article/how-to-hire-a-web-designer**.

If you have already published a book in print and want to offer supplementary materials, like articles or publicity materials, make the information available on your Web site. You can use additional materials to help promote a book that is already in print or that has gone out of print. Amazon offers out-of-print titles when they are available from booksellers, so if you generate enough interest in your title through your blog or Web site, you may see an increase in your eBook sales. If sales spike and there is a demand for your title, talk with the publisher or your agent (if this applies to you) about the possibility of having the book released as a second edition. If the rights still reside with your publisher, see about getting your rights released; once you have the rights, you are free to seek publication elsewhere.

Social networking sites

So why is everyone buzzing about social media? It has considerable value, which is why it is a hot topic, even among Web geeks and gurus. Interactivity is the new trend of the information age. Web sites are increasingly becoming linked and networked, and incorporate elements like video, chat, and forums or message boards where people can communicate. The Internet does what television did for people in the mid-20th century entertainment-wise, and then some. It is the place people want to be since it has so much to offer: research, friends, colleagues, entertainment, information, news, weather, sports, and more. Look at the Internet as the go-to place for the best of professional resources and social connections.

Social networking — that is, chat and instant messaging, forums, newsgroups, and social media sites like Facebook (**www.facebook.com**), Twitter (**www.twitter.com**), MySpace (**www.myspace.com**), and LinkedIn[SM] (**www.linkedin. com**) — are used for both social and professional purposes. Becoming a member of sites like Facebook and Twitter can be simultaneously fun and useful. If you are unfamiliar with Twitter, it is based on a few very simple concepts: network with other people and let them know what you are doing. The longer you are on Twitter and the more people you seek to "Follow," or view their messages (called "tweets"), the more contacts you will make. If you begin following Random House on Twitter, you will be able to see their Tweets on your Twitter homepage, right alongside those of your friends.

Facebook and MySpace are based on the same concept as Twitter, but are more robust. On Facebook (which is surpassing MySpace in popularity and is also better geared toward professionals) you can post photos; write updates on your "wall" (your personal page); link to sites, videos, and music; chat; and private message the "friends" you are connected to. Your friends' updates appear in a sort of queue as they are posted. If you are using Facebook as a professional outlet, you may post updates such as projects you are working on, the book you are writing, writing in general, publishing an eBook, research, any conferences you attend, and so on. You can also join "groups" on Facebook, and create author fan pages and publisher, author, or other special interest groups.

LinkedIn is the social media networking site for professionals. Less informal than Facebook and Twitter, LinkedIn allows you to post a business profile, such as your experience, current positions, memberships, alumni status, awards and accolades, and more. You can connect with other professionals, send direct messages, and "recommend" or be recommended by others you are affiliated with.

Advertising

Some authors think advertising is the way to go, but this drains the pocketbook quickly and can produce relatively little by way of compensation in return. Social media networks are much more powerful and effective ways to get the good word out there; they are also free, and require only the investment of your time. Unless you have a budget for advertising, you may want to try the steps outlined in this chapter before going this route. If you would like to try it anyway, start by seeking advertising space on Web sites related to your book's topic. For example, if you have written a work of political satire, try Slate (**http://slate.com**). Or simply sign up with related organizations. For instance, if you are a mystery writer, go to the Mystery Writers of America Web site (**www. mysterywriters.org**) and spend your money on becoming a member. Your book will be listed on the Web site, provided it meets the organization's guidelines.

CASE STUDY: FOLDED WORD PRESS EDITOR ENJOYS THE KINDLE

J.S. Graustein
Managing Editor, Folded Word
www.foldedword.com
editors@foldedword.com

J.S. Graustein earned a master's in ecology from Northern Illinois University, but switched her focus to writing after starting a family. She writes poetry for children and adults, and those poems have appeared in both print and online publications. She inherited Folded Word, a small indie press, from founding editor Jessie Carty in April 2009. Since then, J.S. has edited and produced Folded Word's first chapbooks (small books or pamphlets) *and book. She also serves as Submissions Chair for Surprise Valley Writers Conference in Cedarville, California.*

As an editor of a start-up indie press, I do just about everything: read submissions, collaborate with writers to revise their works, produce the lit magazines and books in which they will appear, promote the works, and handle all the legal and financial details. I could not do all this without the help of some enthusiastic, talented, and devoted volunteers. Folded Word has some incredible friends.

My first experience publishing for the Kindle, though, was not through Folded Word. My dad writes inspirational romance novels and was looking for a new way to promote his work. So, in Fall 2008, I convinced him to release one of his novellas as an eBook for a nominal charge. I love all things Do-It-Yourself, so I signed up for a DTP account and devoured everything in its "Knowledge Base." My dad sent me his manuscript as a Microsoft Word document and away I went.

But I am a perfectionist. I did not like how the automated tools rendered some of the elements when I did a .doc to .azw conversion, so I decided to produce his book using HTML. While this took longer and required some skill, I far preferred the control it gave me over indents, italics, graphics, and the table of contents.

Uploading the file to my DTP account was straightforward. I definitely appreciated Amazon's ability to add its own identification number rather than requiring an ISBN. Back in 2008, the book went live the same day and we made our first sale that evening. I think it helped that my dad's book belonged to a popular genre. Apparently people with Kindles were desperate for new romance over that Thanksgiving holiday. I was hooked by the whole experience, but I'm not a romance fan. I set out to find a way to repeat it with general fiction and poetry.

Enter Folded Word. When we started our Signature Chapbook Series in Autumn 2009, it was all about handcrafting. The print books are hand-sewn with a hand-calligraphed cover. They sell for a hand-crafted price — which was on the expensive side. I wanted to offer a lower-priced option while still maintaining the hand-crafted concept. I decided to release the chapbooks on Kindle, which, especially for the poetry chapbook, required hand-coding the HTML.

I hope DTP will develop some kind of "stanza" tag soon. Currently, I have to use CSS to define the spacing after lines as well as the indents for wrapped lines. Otherwise, stanzas render as paragraphs and a reader cannot tell which lines are wrapped or where the intended line breaks are. I fear there is very little chance of poetry catching on for eReaders without more publishers taking greater care in their eBook layout. So much of poetry is the rhythm created by the white space. But white space does not translate well on the Kindle; there are too many variables.

Uploading has changed a bit in the last year. It now takes a week for a new title to process and appear for sale. And there are so many more titles for sale now that readers really have to be looking for your title to find it. Long gone are the days of pricing a title for $0.99 and waiting for bargain-hunters to stumble across it and pop it in their carts. Even still, I appreciate Amazon's inclusive attitude toward its digital offerings rather than an exclusive one that would prevent new voices from entering the market.

So far, our printed chapbooks have sold far better than our eBooks. I expect this will continue to be true for many years. The expense needed for a dedicated eReader is a big hurdle for entry into the eBook community. Reading eBooks on a computer monitor is not worth the eye strain or the limitation to mobility. As with all electronics, I expect device prices to drop and ownership to rise — quickly.

I have been a Kindle reader since February 2008 and an iTouch reader since August 2009. My first few trips through airport security with my Kindle were challenging. Amazon should pay me a commission for all the free PR I have done with security screeners and fellow passengers. I quickly learned to treat it like a laptop computer to save time. My iTouch has never opened up conversations with strangers.

Both eReaders give me portability and choice when I am out and about with time to spare. I loathe waiting with nothing to do, but dragging along magazines and books to the grocery store or the school pick-up line is not convenient or pleasant. When compared to my iTouch, my Kindle has a larger screen, no backlight, and generates larger and crisper fonts. I also like the weight and the bulk of it in my hands. It feels like I'm truly holding something.

In December 2009, I expanded my Kindle reading from published eBooks to submitted manuscripts. It was fantastic to handle a stack of chapbook submissions without worrying about losing pages or burning up my eyes on the computer screen. While the formatting was crazy since I used the free automated conversion service, the convenience and comfort of reading manuscripts on my Kindle made handling submissions a joy.

But as a reader, I prefer having the choice of medium rather than preferring one medium to another. I travel frequently and suffer from migraines. In both these circumstances, I far prefer to read on my Kindle. I can pack a slew of books to suit all my moods, plus references for my writing, in my purse. My Kindle has no backlight and does not flicker, so it is just as pleasant to read during a migraine as a print book. Unlike print books, however, I can turn up the font size so I can focus on the letters and read without effort.

I definitely feel an extra surge of excitement when one of my poems makes it into print. There's something extraordinary about a publisher thinking my work is worthy of the physical and monetary resources print requires. But I also love seeing my work digitally displayed because I know that many of my friends will read it if it is just a click away. I have not had my own writing published on a Kindle yet, though I think it would be fun to see my name attached to something for sale on Amazon.

I think today's digital books are in a transitional state. They have pros and cons, benefits and drawbacks. But they are a necessary step toward the development of digital content that will keep future generations connected to literature. I am sure there will always be a need for print books. Print books will always be treasured and passionate people will always find ways to produce them.

Today's children are developing in a digital age. They see interacting with digital devices as a natural part of existence □ as natural as breathing, eating, and sleeping. Whether or not that is a step forward in human development is up for debate. But for books to survive, they'll have to be presented in a way that's relevant for these new readers. Today's digital books are a step toward that relevance.

The biggest challenge eReaders pose is one of perception. For as long as people expect eReaders to give them a print experience, there will always be disappointment and criticism. It is like eating a pear while desiring the experience of a banana.

It will be a long time before it will be possible to render complex and artistic print layouts on an eReader. Poetry will forever be a stumbling block, since different people will read at different font sizes, limiting any control over layout and visual elements. And until the last analog generation passes on to the great beyond, eBooks will be seen as yet another electronic threat to traditional art.

Yet eBook publishers can meet the perception challenge with some ingenuity. eReaders have functionality that print books don't. This has the potential of opening up new genres and new modes of storytelling. But it will take some experimental writing and publishing to exploit these functions. If writers and publishers do rise to this challenge, however, people may decide that it's not a question of eBooks being better or worse than print books; just different.

Reviews

Reviews of your work are some of your best allies when it comes to promoting and selling copies of your eBook. Ideally, before writing a single word intended for publication, you will have read widely in your area of interest and will be familiar enough with positive and negative criticism to have a solid understanding of what are considered weaknesses and strong points in a work — whether it is fiction, nonfiction, or poetry. The best time to send out review queries is three to four months before the book is scheduled to print. In your case, three to four months before your work is scheduled to go live on Amazon. The industry standard is to send out query letters for review just as you would when seeking publication from many traditional publishers.

Book reviews and publicity

Reviews are a powerful marketing tool for authors, who can use the copy for promotional purposes. Reviews provided before the book's release can be used to excerpt blurbs from (the quoted promotional copy that appears

CHAPTER 9:
Reviews and Managing Published Work

on the back jacket and sometimes in the front matter, or beginning pages, of books), and for many other purposes, such as for excerpting on an author's blog, Amazon Author Page, or personal Web site. Authors may select to include (positive) review copy to appear with their listing on the Amazon, Powell's Books, or Barnes & Noble Web sites, and may quote portions of a review on the back cover. Of course, a review is copyrighted material just like any other published printed matter, so use excerpts sparingly and always be sure to include proper attribution.

If yours is a nonfiction book with a historical subject, send review queries to publications that review nonfiction as well as those concerned with the subject matter, such as historical societies and organizations. Send out review query letters and mail a hard copy (or e-mail an electronic copy) to those who respond positively to the query letter rather than blindly sending copies of your book out to publishers and organizations that may or may not provide a review. Some reviewers will not review eBooks, POD editions, or self-published works; it is a best practice to view submission guidelines before sending in your work. *See Appendix D for a sample review query letter and Appendix E for a sample review slip (this is included with any editions of your book that you send out for review).*

Be aware that there are different types of reviews. They are:

- **Prepublication reviews**: Reviews that release before the book. Written for those in the book industry, such as wholesalers, bookstores, and libraries. These reviewers expect to receive bound galleys. A galley is a prepublication book copy that is usually printed in black and white.

- **Early reviews**: These are packaged with a review slip, press release, copies of other reviews, and other materials, like brochures, that add credibility to your book and convince reviewers of your book's credibility. *See Appendices C, D, and E for a sample press release, review query, and review slip, respectively.*

- **Post-publication reviews**: These are intended for the consumer and appear after the book is published and, usually, available in bookstores and retail outlets. In your case, of course, post-publication reviews will often come in the form of reviews on Amazon, which appear on your book details page.

For major review outlets like *Publishers Weekly*, who review fiction, nonfiction, mystery, children's, mass market, and other select titles (but no self-published material; see **www.publishersweekly.com/article/CA6428088.html** for complete guidelines), send a press release with a cover letter and a bound galley three to four months (four is preferred) before publication. Follow up one to two months later by sending them a final copy of the book. It is important to send the galley first since *Publishers Weekly* is a prepublication reviewer. Your book may not be reviewed, but if it is, it can mean a major increase in sales to the *PW* market, and of course, any print sales of your book can translate to sales of your eBook edition. Follow up with a nice letter or e-mail to anyone who takes the time to review your book, and let them know you appreciate it, even if the review is negative. Be professional, and steer clear of tongue-in-cheek comments. Remember that anyone who reviews your book has put time into reading it, and most likely was not paid for doing so, unless they work for a major review outlet like *Publishers Weekly*. A very professional site that self-published authors might consider is Self-Publishing Review (**www.selfpublishingreview.com**).

Book reviews are one of the best forms of publicity you can receive as an author, along with interviews and articles.

Encouraging reviews of your work for Amazon

There are several things you can do to create awareness of your work besides the marketing and promotional methods mentioned in chapters 7 and 8. Encourage reviews by reviewing other eBooks. Of course, avoid "reviewing" anything you have not actually read. In general, reviews that you post will be visible for both the print and eBook versions of a work, even if the two versions do not share an ISBN or ASIN (Amazon Standard Identification Number). For example, a comparison of reviews for Michael Crichton's *Pirate Latitudes* in both its print and electronic versions reveals the same results, even though they are not linked numerically (the print version has an ISBN-10 and ISBN-13, and the Kindle edition has an ASIN). So with that in mind, any review you post will be visible for all versions of the title. But there is a catch — at the end of your review, post a signature that includes your name and the title of your book. This simple tactic, when professionally done, can be effective.

Look at the reviews for other books, both print and electronic, that are similar to yours. If you have written a historical thriller, browse through the reviews for other historical thrillers. Look for reviews that are thoughtful and well-written, and consider sending a review query to the people who wrote the review. Looking at the name associated with each review, you might notice that beneath the name a reviewer status may or may not appear. A status badge can include:

- **Real Name**: This means that the individual's name has been verified from a credit card.

- **The word "the"**: A verification of celebrity status. For instance, if you saw "The" beneath author Margaret Atwood's name, you would know it was really Margaret Atwood who wrote the review or comment, and not someone using her name and pretending to be her.

- **Community Forum '09**: This person participated in the 2009 Amazon Community Forum at the Seattle, Washington corporate headquarters.

- **#1 Reviewer, Top 10 Reviewer, Top 50 Reviewer, Top 500 Reviewer, Top 1,000 Reviewer**: These are badges that indicate reviewer status.

Badges are obviously indicators of status, and active reviewers might be the most interested in receiving a query from you and a free review copy. To get a reviewer's e-mail address, click on his or her name to see the user profile. Often, a Web page or e-mail address will be listed. You may have to do some sleuth work, but it can pay off. Of course, be professional and courteous. Even though you are e-mailing your request, send a query letter (*see Appendix D for a sample*) in the body of the e-mail and ask if the reviewer would be interested in reviewing your work for Amazon before making the assumption and sending your work.

Of course, if any readers send you positive feedback on your material, ask them if they would kindly post their thoughts as a review. You can ask the same of willing friends, family, and colleagues. Many of them will be happy to support your endeavors. Some Kindle authors include an e-mail address in their ePublications encouraging readers to send — depending on their subject matter — tips, helpful hints, and suggestions. When you respond to e-mails, include a subtle yet gently encouraging postscript at the end of your response asking readers to post a review if they enjoyed your work.

Monitoring reviews and how to treat good and bad

What is the best way to find anything on the Internet? A Search engine, of course. Google is arguably the best one out there; it remains in the lead among its competitors, and certainly is a helpful resource. To stay on top of the online buzz, conduct a Google search of your name from time to time. While seeing what the Internet dredges up on your name may be unpleasant, it is a necessary task. As an author, you are bound to be mentioned eventually by someone, unless you do no marketing whatsoever and do the online equivalent of hiding under a rock (no blog, no Web site, no Twitter or Facebook marketing, no word of mouth marketing, and so on). *You can use Google Analytics and some of the other methods for monitoring your online reputation, discussed further in Chapter 11.*

On Amazon

If your work is available as a Kindle edition only (you have nothing in print), the majority of your reviews will come in the form of customer feedback on your book details page.

The best method of review monitoring is to get in the habit of checking your reviews regularly. How often you choose to check your reviews will largely depend on how many copies your eBook is selling. If you have only sold a handful in several weeks, you are safe checking reviews once or twice per week. If digital copies of your eBook are flying off the virtual shelves, it would be wise to check on reviews daily.

While you cannot remove negative reviews, you have the option of saying whether or not the review was helpful (by clicking the "Yes" and "No" buttons beneath the customer review. You can also click the "Report this" link, which is also situated beneath the review. Clicking to report a review allows you to report the content as inappropriate. If Amazon deems fit, they will remove it. Even before any negative reviews are posted, try being proactive by following the suggestions detailed above in the "Encouraging reviews of your work for Amazon" section. Seek positive reviews, and if any negative reviews come in, seek even more positive reviews. If you only have a very few reviews, the negative ones will carry more weight; this is never more evident than when they bring a 5-star rating down to three stars or lower.

Other Web sites and in print publications

Most of the top-name reviewers, like *Kirkus* Reviews, ignore eBooks altogether, or anything that is the result of a POD printer or self-publishing effort. *Publishers Weekly* has many hundreds of books come to them in a given week, and carefully select the books they wish to review. They ask that, with each submission sent to them, the query letter include a book synopsis as well as publicity information, such as whether there are book clubs, paperbacks, audio rights, movie sales, and advertising and promotional budgets more than $30,000. It is a no-brainer to guess which book would be selected, given a choice between the unknown author making her debut and the author with a contract to write three more books for Random House whose most recent has just been optioned for movie rights.

But if you have published a print book and are now going forward with an eBook attention, kudos to you. Send in your books to all the usual channels, and use any reviews that come in for promotional purposes. Wherever possible (on your blog or Web site, for instance), link to your book detail page for both your Kindle and print editions. Self-published authors, however, will have a bit of a harder time securing any print or Web reviews. There are many good reasons for this, but your chances are improved the better your work is. Authors who are published with small and independent presses probably have the widest outlet for reviews; scores of small- to mid-sized journals and literary magazines review such works all the time. There are really too many to mention, but some such reviewers are listed on the New Pages Web site, at **www.newpages.com/npguides/reviews.htm**.

Tracking Sales Consistency

Sales rank is a number that measures how well a product is doing overall in the Amazon "catalog." The number is only available to the person (publisher, customer, author) who posted the product for sale. The category sales rank reflects how well a product is doing in its category. Sales rank is updated hourly.

Keeping an eye on sales and sales rank is good for a few reasons. Of course, it is a good idea to monitor sales so you can step up your marketing and promotional efforts when sales begin to lag, but since sales rank is updated so frequently, you can also use it to gauge the effectiveness of marketing campaigns.

Say you launch a marketing campaign for Twitter, for example. Your campaign is simple: This week, you plan to tweet one interesting fact from your book every other hour, and provide the link to your book page. Next week, you plan to start a Facebook campaign. You already have an author page and have gained 500 followers, so now you will post one status update per day that includes, as with your Twitter campaign, a helpful or interesting fact from your book or article along with a link to your book page. You will monitor the effect this has on your sales. At the end of the Facebook campaign, you will compare the sales you had for each week. Which was more effective? How effective would a campaign launched for a week on both Facebook and Twitter be?

Keeping Published Content Current

Like with a Web site, it is vital to keep content current. If you have read any degree of articles or newsletters about Web design, SEO, or Internet marketing practices, you may have come across the phrase "Content is king." Indeed it is, and so is currency (currency in the sense of how current your keep your material, not money, although this type of currency can certainly lead to money). Keeping abreast of changes related to your topic is vitally important, especially if you have published a nonfiction title. The moment any information becomes outdated or is otherwise in need of an update, you should post the changes to your document and re-upload it to DTP (and Mobipocket).

Keeping your content relevant will keep your eBook from becoming outmoded. Look at the fact that you can update your book at all as a leg up on traditional print publishing. Even if your material is already published in print format, you can release updated editions (be sure to specify whenever your book is subsequent updated by denoting second, or third editions, and beyond). Traditionally published works of nonfiction beg for another writer to pen an updated edition as soon as any information in yours is no longer current.

CASE STUDY: YOUR TOP REASONS FOR PUBLISHING WITH KINDLE

Web site: www.thecreativepenn.com
www.twitter.com/thecreativepenn
www.facebook.com/thecreativepenn

*Joanna Penn is the author of three books and is a blogger at **www.TheCreativePenn.com**, which is one of the Top 5 Australian Writer's Blogs. She is also a speaker and international business consultant. Her books about writing, publishing, and book promotion are all available on the Kindle. You can get a free Author 2.0 Blueprint on using web 2.0 tools at **http://author2zero. com**.*

It is definitely time for authors to embrace technology and devices like the Kindle – here's why.

- **People are online and searching for content on the Kindle and other devices**. Those people could buy your book. Even if you do not like consuming eBooks, millions of other people do and more start using the technology every day. You want to reach them, so your books need to be available. If your book is not available in e-format, you may miss a sale, as many e-users will not buy print books now. As the Kindle and other devices continue to improve, more content will be provide; you do not want to lose out on these potential sales.

- **EBooks are global and immediate**. Has your print book been distributed to all bookshops in all countries where the Kindle is sold? It is extremely unlikely. Your book can be truly international when it is a digital version. People can also buy as soon as they hear about your book through the wireless Kindle Store, instead of waiting until they reach a book shop and then forgetting about it.

- **Find new readers and sell more books**. People can download samples on the Kindle — so they are more likely to try out new authors that they would not initially buy. In addition, once a Kindle user finds an author they like, there are links to more of their books, making it easier for people to buy multiple books. This is a real bonus, and you will often see the top Kindle categories have multiple books by the same author because of this feature.

- **Make more money per book**. If you are self-publishing, you will likely make more money from an eBook sale than from a print book sale. There is no printing or shipping cost; so all income is profit after Amazon/other distributors takes their cut. This post by popular author Joe Konrath goes into financials of selling on the Kindle with and without a publisher **http://jakon-rath.blogspot.com/2009/10/kindle-numbers-traditional-publishing.html**. His books sell enough on the Kindle to pay his mortgage.

- **Update books easily**. For non-fiction authors, you can update an eBook very easily — just load another version. This means time-sensitive information can be updated without reprints for no cost.

- **Test the market**. If you have a book but do not know how the print response will be, test it as an eBook first and see how it performs. Then you will have more of an idea whether to publish in print also. I think publishers will start using this model, as it is very low-risk and low-cost.

- **Sell your backlist and other information**. There is no minimal page count to the 'books' sold on the Kindle. Therefore, you can sell your backlist and include short stories, old books that have gone out of contract or print, essays, and other information. These may provide extra revenue as well as a way for people to find your more recent books.

- **Why not have an eBook version?** The tools are free and easy. You do not need to know much to publish on the Kindle yourself. If you have a basic text-based book, it is a simple conversion to HTML and upload onto the Kindle Store at **https://dtp.amazon.com**. Set the price you want and you can start selling immediately. You can also just load a Word document to **www.smashwords.com** and have the file converted and distributed for free to the Kindle as well as Sony Reader, Barnes & Noble, and more. There is no reason not to have an eBook version of your book these days. If you are really technophobic, use a service like **www.ebookarchitects.com** to convert books for you.

Sparking Ideas

Some writers have no trouble generating ideas for new content. You may know them by their workspaces, which are often brimming with notebooks, folders, and shoe-boxes bursting with scraps of paper — or maybe their computers contain the digital equivalent of overstuffed notebooks. Maybe you are not such a writer, or you may just be experiencing a dry spell (not to be confused with writer's block, which, according to some researchers, stems from a temporary disconnection between the part of your brain that edits and organizes the writing and the part that controls and processes ideas. Think left brain versus right brain). This section discusses techniques you can try to start ideas flowing. They are intended as a starting point, rather than a one-size-fits-all approach, so as you read through them consider which ones might work for you. Or, if you have no trouble in this area, you can safely skip to the end of the chapter for tips on writing a series and publishing short-form content for the Kindle.

CHAPTER 10:
Generating Ideas for New Content

On the other hand, if you are plagued by so many ideas that none of them ever make it to a coherent or finished form, brainstorming may be a good option. At first, you might need to force yourself to write down as many ideas as come to mind and then organize your thoughts later. Sometimes the hardest thing is to just focus on a single idea. Often, it is most difficult to sit down and actually complete what needs to be done in order to accomplish anything. Writing takes discipline, so be sure you set aside time to work on your ideas, whether you generate several of them or you single one out and begin working toward a rough draft.

Mining the Best-Seller List

What better way to find out what is selling than to visit the best sellers list in the Kindle Store? Other places to look for ideas, if a best seller is on your agenda, are the *New York Times* best-seller list, anything that *Publishers Weekly* or *Kirkus Reviews* gives rave reviews to, and the regular Amazon bookstore. As you are looking for ideas to write new material, read widely and often in your genre. Read best sellers and literary fiction, if fiction is your forte. But do not stop with the best-seller lists.

It is worth keeping in mind that books that are currently in the forefront of public awareness do not necessarily equate good writing. But if your only concern is selling your material and not achieving a durable, lasting quality to your work, then best-selling fiction is all you need concern yourself with. Because publishing is controlled now by six media mega-conglomerates, the current bottom line in publishing is what will sell out of the gate, not what will contribute to great literature. If you are less concerned with hitting the best-seller list and more involved in the depth and quality of your writing, read even more widely, and more often. There are many independent publishers out there who offer high quality writing and whose primary concern is not blockbuster success for every book printed.

Current Trends, Fads, and Hot Topics

Paying attention to current trends and fads can be important for a few reasons. First, you should avoid any topics that have been overwritten or that have been written about so much that they are entering the realm of the cliché. Second, remaining attentive to the writing current can help you tap into what might sell. Or, looking at the same idea from a different angle, it can help you avoid trendiness for its own sake and push you more toward originality. My own recommendation would be to take the latter route, but of course not every writer has the same concerns or agenda. The most important considerations are to remain aware of what kind of writing is out there by becoming a diligent reader and Internet researcher and to remain true to your own writing. Nurture your own style by giving your writing the time and attention it deserves, and if you have not already, you will find your niche and your voice will come across as more authentic to your readers.

If you are a nonfiction writer, pay close attention to the topics that are getting at least a fair amount of press time and what is current in your area of expertise. Are you a technology writer? If so, your work is cut out for you; technology is in a constant state of evolution and development and there is always some new software version or gadget to write about. But whatever your topic is — including domestic affairs, travel, or global economics — if you are searching for what to write about next, you would do well to start by scouring the national news outlets online. Read the latest books in your field and as long as you stay on top of the most recent developments in your industry, you should have no trouble generating material for your next book.

Journaling and Observation

Journal writing is a way to stimulate ideas without feeling pressured to come up with anything specific. Writing in a journal also allows you the pleasure of writing just for the sake of writing. There is no structure, limitation, or deadline involved. It can be a great way to tap into ideas you did not realize were there. Use a journal to write down your observations about the things, places, people, and events around you. You can also use the journal to determine what time of day your writing is best or when you are feeling the most productive. Experiment by writing at various times of the day and night and see when you are at your most inspired or prolific. Try not to read or judge what you have written until you have finished journaling for that day or until the next time you write in your journal.

Free-form journaling involves writing about whatever comes to mind:

- Write about what is going on in your life, your concerns, thoughts and emotions, and anything else that moves you, such as memorable dreams you have had.

- Look around you and write about what you see.

- Try thinking of familiar objects in new ways by taking a fresh look at things you see every day.

- Could any of the objects around you be used in a story? Look for details.

- Observe people around you; try taking a walk or visiting a public place and looking at the activities people are engaged in. What are they doing? What will they be doing next? What are their lives like? The answers to these questions may provide material for your characters or next story line.

- Try structured journaling, where you restrict your writing to a single topic or idea.

- Try writing about a different topic every day, or keeping a journal dedicated completely to writing about cooking, sports, dreams, or anything else you are interested in.

- Try keeping a journal dedicated solely to writing down observations.

- Visit different places — go to the park, the mall, or take your journal with you on a city bus and write about what you observe.

- Try writing down whatever fragments of conversation you can decipher from the chatter around you. When you come back to them, you may find inspiration in the ideas from other people's conversations.

- When you re-read old journal entries, you may even notice how far you have come in your writing a year or so down the road.

With journal writing, you can keep track of experiences that inspire you or that are simply interesting. Anything out of the ordinary, anything that strikes a chord, and anything that is interesting, odd, or unusual is fair game. You may choose to keep a separate journal of interesting experiences or have an area in which you can record thoughts on stories you read, ideas for new tales based on stories you hear other people tell, events from your own past or present, or simply enjoyable moments with your children, relatives, and friends. If you are struggling to remember an event from the past, write what you remember. Sometimes this technique can trigger more memories, but the beauty of fiction writing is that you can take many liberties to alter facts and include new information that moves the story along, enhances the character dynamic, and increases the overall interest of the story for your readers.

If you are a nonfiction writer and you are struggling to piece together snippets of memory, start by writing about what you do remember. Then talk to others who were witness to the event in some way, but keep in mind that even if it is nonfiction, your story is just that: your story, told from your point of view, written in your words. Historical nonfiction is a bit different from personal writing, of course, and not just in the way you assemble information, but also in the way you approach the writing style. Unless you are writing an autobiography or a biography of someone you know (which is fairly rare), your information should almost always come from primary and secondary sources on your subject.

Life Experience

You have probably heard it said that truth is stranger than fiction. Real experiences can make for some of the most gripping story lines and uncanny sources of inspiration. But keep in mind that details of people, mood, events, and places can also be — and sometimes should be — changed. The key is to listen to the story. If you experienced a traumatic car accident in which no one was harmed and you think the experience would make great fodder for short story or novel, you may be right. However, as you are writing the plot and planning your characters, consider what would make that true event more interesting. What if someone did get hurt in the accident?

Sometimes reconsidering true events for a story takes a reassessment when it comes to the theme and audience. Think of how you could adjust your story to make it more interesting to the reader. Remember that if you were writing about a true event, it would be nonfiction if you changed none of the details; memoirs are a little different, and artists can take a little more license since a memoir is essentially the story as the person writing it recalls it. However, for fiction, let true events inspire the work rather than enslave it. In the example of the car accident, you might ask yourself several things: What if? What if, instead of everyone arriving home shaken, but safe, someone had to be rushed to the hospital by ambulance? What if the action were heightened with the victim learning she is pregnant? How could you heighten the interest of the situation for your audience in particular (or who you can best determine to be your audience)? Look for ways to make the reader feel included in the situation, and the result will be a pair of eyes that is glued to the page.

Looking to the Media

Stories are all around you. Television, magazines, newspapers, and the Internet: all are overflowing with information, details, narratives, and general interest content. A child saves his mother's life by calling 911; a young entrepreneur starts a charity and makes a measurable difference; a female scientist makes a groundbreaking discovery. News items and articles abound with story potential, even if they are about famous people, politicians, and other public figures. Think outside the box. What if you substitute the public figure in a news article for that character you have been germinating in your brain? What if you used the essential plot of a story or article and imagined the rest of it? What would be your characters' roles in the events? Remain open to other forms of media, such as advertising, music, posters, e-mails, Web sites, and movies, as these can sometimes spark unexpected ideas.

The media can also clue you in to the hot topics of the day, as mentioned earlier. Most readers are interested in current trends and events, and many works of fiction (not to mention films) have resulted from imagined crises similar to true events. Consider looking for ideas on historical references or current events that hold potential interest, but perhaps are lacking in information — this is a solid technique for both fiction and nonfiction writers alike. As a fiction writer, you could create a story based around the possibilities of location, people, places, and events surrounding a historical site, event, or discovery. If you are a nonfiction writer, you are afforded plenty of room for investigation and research.

Look for ideas and stories that have not received adequate coverage in your genre. Can you fill a gap in the literature, or present a new angle on a semi-researched topic if you are a nonfiction writer? Perhaps current findings and research are needed to update the topic. Before committing to a topic, ask yourself whether it would be of interest to your audience, if it would have a perceived impact on the lives of readers, how much information is available, and what information the current literature provides.

Brainstorming and Mind-Mapping

Brainstorming is the ultimate form of free-association writing. Brainstorming involves generating as many ideas as you can about a single topic or about as many topics as venture into your mind. Anything goes. Try starting with a single idea, topic, place, or event, and enjoy the randomness that ensues. You can brainstorm with a friend, a child, or anyone who is willing. The exercise can be fun and you may choose to write the results in your journal when you are through and have processed your ideas, or jot them down as you go along. If you work best under structure, try setting a timer for 15 minutes and brainstorming as many ideas as you can during that time. You may not want to stop when the timer is up. Written material can come in the form of word lists, prose, or mind maps.

Mind maps typically originate with a single idea, topic, place, or event, written in a circle in the center of a piece of paper. Around that word, draw other ideas that branch off from it. For example, you may start with the word "catastrophe." This word is written in the center, and is the main or starting idea. Lines may be drawn out around the center circle to other words, like "building," "foreign affairs," "international intrigue," and anything else that comes to mind during your brainstorming session. Make as many mind maps as creativity allows, or get even more inventive and create a large-scale mind map on a bigger piece of paper such as newsprint or poster board.

With your mind map, word list, or journal entry complete, look over the results. Are there any ideas or words in particular that spark your interest? Perhaps you see the makings of a story or even a rough plot in the details. Ask yourself "what if?" as you go along. What if certain events are substituted for others? What if any elements on the paper were changed? How would it affect the story? Try transcribing your ideas into prose form in your writing journal. This may spark other ideas or possibly develop into a rough draft.

Getting Ideas Down on Paper

Sometimes snagging those great ideas means keeping a notebook on the night stand, and maybe an additional one strategically placed in your office. Sometimes it means keeping a notebook in your purse or pocket or otherwise handy wherever you go. If you are the type of writer who tends to get ideas more or less at random, then make sure you write down your thoughts before they disappear as quickly as they arrived. Having a single place in which to write ideas is especially handy when it comes time to look through them for a story idea. On the other hand, if you tend to be more organized and prefer to mull ideas around for days or weeks before writing them down, then you may not see a use for a notebook.

You may find you already have notebooks stuffed full of ideas that have never had the chance to become the stories they want to be. Perhaps they will never amount to anything more than a single sentence, or maybe they will not end up as part of anything at all. You will never know until you try them out. Give your ideas a chance and see how they develop — they may surprise you, or you may surprise yourself. If you are already working on a story, scour the notebook to see if anything is usable and if it sparks anything new. Ideas have a way of losing their magic if they sit too long unattended or if you take too long to write them down. However, if you have more ideas than you can work on in a reasonable amount of time, losing the spark may be inevitable. One way to counteract this is to be as specific as possible when writing down an idea to ensure you retain most of the spirit. When you get around to working with it, you will have a clearer picture of where to begin.

Working with your ideas — rather than just letting them accumulate — takes commitment and dedication. Make it a point to have a certain day of the week where you work with at least one new idea during a designated writing time. If an idea does not work after you have given it more than its fair chance, discard it and move on. Be open to changing the idea; sometimes it is useful because it leads to new inspiration or other ideas that can turn into stories, plots, or characters. Do not feel too tied to an idea. Let it be messy at first, but by all means, get it down on paper. Too often potential stories are never fully realized because of the common misconception that "real" writers simply sit down and beautiful words flow from their fingertips to form an instantly complete story.

The reality is that writing often starts out messy, with inconsistent characters and plot lines full of holes. Some of the best stories are what they are because of the amount of revision they have gone through. There are exceptions, but they are just that — exceptions. Good writing takes time, effort, and a lot of work. Most likely, you already know this if you have completed a manuscript or even a demanding short-form piece.

Writing Best Practices

When you start to write, do it without stopping. Usually an internal editor will want to take over and re-read each sentence or paragraph as it is being written. Save the editing for last. It takes practice to write without stopping to backtrack, but it can be accomplished in small doses at first. Try writing for five minutes without stopping to look over what you have written. This is one way to start ideas flowing, and it can be a good way to begin every writing session until you get the hang of it and develop your own routine. You can try using a different writing prompt each day to begin your writing sessions, such as writing from the point of view of a person you admire, imagining what it would be like to have only a week left to live, or imagining that you have invented something that will change the direction of everyday life or technology forever.

Once you are working on a manuscript, you may find that you no longer need these techniques. You may want to come back to writing prompts every now and then to get the creative juices flowing. Many writers use pre-writing activities on a regular basis. Pre-writing is anything you do to help the writing along, such as character outlines, brainstorming, and note taking. It is often helpful (and recommended) to know your characters, your plot, and your overall story very well before writing a word. Of course, those elements are always open to change in later drafts.

Many hopeful writers see their ideas go nowhere, or result in partially finished stories, because they have too many ideas at once or become unnecessarily stunted by one aspect of the story; maybe the character is not fitting together quite right, or more research is needed, or the ending is becoming a stumper. Make note of where you are stuck and move on. You can come back to it later, but do not let one troubled spot bring the entire story to a halt. This is where a writer's group or feedback from someone who knows about writing can come in handy.

Getting into the habit of writing every day is good practice, even if it is only for 15 minutes at a time. At first, you may spend most or all of your writing time on warm-up or practice exercises. That is fine, since the most important thing is to write and develop a sense of habit and routine. If you do not come up with anything usable for a while, make a fresh start every session and simply discard old writing you are unhappy with without dwelling on it too much. Try to maintain your concentration while you are writing and keep your focus for the duration of the time. Remove all distractions from your work area. If you work better with noise, turn on the radio (but turn the phone off). If your area is too noisy and you need quiet to concentrate, try wearing earplugs and turning on a fan to create white noise and block out distractions. Do whatever it takes to feel comfortable and relaxed in your writing environment. Experiment with different techniques to find what works best for you. And do not forget to take "brain breaks." Stretching for a few minutes when you feel yourself starting to lose focus or getting a glass of water can help you to re-focus when you return to your writing.

Short Form

Do not discount the potential of short-form work for the Kindle. The same appeal of brevity applies to material published in electronic form as it does for anything in print: shorter pieces can be read faster and hold appeal to someone who does not have time to read a longer work. And of course, if someone is looking for an article or short story to begin with, the Kindle holds the appeal of convenience and low price. Price your short-form piece to sell. Depending partly on length and usefulness, shorter works may run from about 99 cents to about $2.99.

Articles and short stories are two of the most obvious examples of short-form work. But some authors also choose to publish excerpts from longer works. This can be useful if you are looking to provide a lower-priced sample of your work, which can be an excerpt from a nonfiction book, or a stand-alone chapter from a novel, for example. If the reader likes what you have to offer, she may be interested in buying the entire work. It can also generate interest in a work-in-progress and in other material you may have available, both on the Kindle and in print. For this reason and for general promotional purposes, it is always a good idea to include the links to your blog, Web site, or other material online in every item you publish. Make it easy for people to access your other work and to find out information about you as an author and professional.

Writing a Series

Speaking of short-form content, another option to consider is serialization. This may be a solid choice if you have already written a longer work — such as a novel or nonfiction book — or if you are or will be generating material on a regular basis. If you choose to publish excerpts from an already-completed longer work, it may go without saying that each one should stand on its own. If you publish a chapter from a novel, for example, that chapter should have enough of a beginning, middle, and end to satisfy the average reader — which will also encourage future sales and word of mouth publicity.

You should also provide a teaser at the end of each chapter or excerpt that will make the reader want to purchase the next installment; often this is included in the writing as part of the story line. But for nonfiction especially, this can be as simple as adding a sentence such as, "Don't miss the next installment of *The Life and Trials of the Presidents*, Chapter 4, featuring James Madison." Make sure when publishing anything that is part of a larger body of work that information about the chapter or number in the series is presented up front, both in the product description and within the first several pages of the excerpt. If someone thinks he is purchasing an entire novel or book and instead gets only an excerpt, you can prepare for many returns.

Whatever you are publishing, be sure to get material out on a regular basis; adhering to a predictable schedule is especially important if you are publishing a work in progress. And of course, as with anything you publish for the Kindle, provide links to your Web site, blog, or other professional site. You may even choose to include a contact e-mail for readers to provide feedback. Make it as easy as possible for people to find information about (and purchase) other books or articles you may have written. Sales on the Kindle have a way of building on themselves; the more work you have available for purchase on the Kindle, the more you will sell. And the more you sell, the more your product visibility is increased. With the right marketing, attention to quality in your work, and positive reviews, your work could well find its way onto a best-seller list.

CASE STUDY: ROMANCE AUTHOR KICK-STARTED CAREER WITH EPUBLISHING

E-mail: lynneconnollyuk@yahoo.co.uk
Web site: www.lynneconnolly.com

CLASSIFIED CASE STUDIES
directly from the experts

Lynne Connolly writes paranormal romance and historical romance for the three largest ePublishers: Ellora's Cave, Loose-Id, and Samhain Publishing. Her books are available from the publisher and in Kindle format, in e-format and in some cases, also in print. Lynne lives in the UK, near Manchester, with her family and her mews, a cat called Jack.

How did I break into the traditional print publishing world? I never did. I started in ePublishing in 2000, when it was a very small industry, and now my publishers bring my eBooks out in print. It has now grown so much that I am comfortable where I am.

I wrote a historical romance, *Yorkshire*, that got a lot of "we like it but we don't think we can sell it" responses from the UK press — enough to make me search for somewhere that might take it. At the time, we had just bought our first computer, and so I hooked up the phone to the Internet and went looking. I found writing groups first, and met some people online who helped me to polish the manuscript, then I was recommended to an ePublisher that no longer exists. *Yorkshire* is now in its third, and I hope final, home at Samhain. Each time, it has been rewritten and re-edited.

I was very keen to have my work made available on the Kindle, but that was down to my publishers. Ellora's Cave, Samhain, and Loose-Id have all signed agreements with Amazon, and last year, "Yorkshire" was offered as a free read. I was delighted with the result, as it gave a kick to the rest of the Richard and Rose series, and the rest of my backlist at Samhain. I do not deal directly with Amazon — my publishers do that. The Ellora's Cave deal is a little different to the others, as Ellora's Cave was keen to preserve the royalty rate it paid to its authors, so they have factored in the cost of selling through Amazon. Consequently, the books are cheaper on the Ellora's Cave site but still in the affordable range for Kindle readers.

My books have always come out in eBook first. Sales have increased exponentially so that now I am earning what a mid-list, traditionally printed author in New York would earn. I do have an agent, but because of the cushion of my eBook earnings, the fan base I have, and my reputation, we can take our time looking for the right print deal with the right publisher. Ebooks have made me marketable with a readership and reputation. I still sell more books from the publisher's Web site than anywhere else (including Amazon) but third-party distribution is becoming more important to sales.

From a financial point of view, ePublishers work on the royalty-only model. Royalties are much higher than for print books (between 30 and 50 percent of cover price) and they are paid faster — typically a month or two months after the sales. There are few returns to factor in. This is completely different to the print model, where an advance is paid, few books are earning out, and when they do the royalty is lower. An ideal would be a mix of the two, and many authors are now choosing to remain in both markets. A regular monthly check plus the boost of the advance payment works well for many.

It also means that everything is done electronically, from submission to editing to release. That means that although I live in the UK, with my market mainly in the United States, I can be in daily contact with my readers and editors. It means I can compete on an equal basis with my U.S. compatriots and it is a big reason why I chose to remain in the ePublishing world. I do travel to the United States every year for one of the big conventions, so I can meet readers, editors, and colleagues, but it is not compulsory.

My books are available worldwide. Territorial rights mean little to ePublishing, as once it is on the Internet, it is available in almost every part of the world. I have had fan letters from Africa, India, and Australia, which I find very exciting.

Keeping a Finger on the Collective Pulse

In the Web-centric, technologically advanced age we currently enjoy, information from nearly every conceivable industry and field is readily available online. This makes it easier to both suffer from the maddening yet elusive thing known as information overload, and to stay up to date with the latest developments of a fast-paced, increasingly global society. Writers, especially, have an obligation to stay current. But where to begin? Start by making information-gathering part of your regular routine by setting aside a certain amount of time every day — just like you would for writing or other work — for reading online articles on topics in your industry.

Make your time spent reading really count. Employ a methodical perusal of carefully selected sites, such as ePublishing and Kindle-focused blogs and Web sites, as well as Web sites on publishing in general. *See Appendix A for site lists*. Over time — through links, reading, and references — you will discover new sites that interest you and will most likely develop a habit of visiting certain

CHAPTER 11:
Up to Date: Staying Abreast of Changes in the Industry

places online over and over again. Bookmark what you find useful. Take note of the links provided in the Appendix — a list that is by no means intended to be exhaustive, but that can get you pointed in the right direction — and the list of social networking tools further on in this chapter. Besides reading Web sites, take advantage of various alerts, applications, and forums that can help you stay current.

Making Google Work For You

Google Alerts

Speaking of staying on top of industry trends, signing up for Google Alerts allows you to specify keywords and key phrases and to receive e-mail notifications any time those topics come up in search. Visit **www.google.com/alerts** to sign up. As an author, you might want to enter your name and the title of your book in the search terms. Separate each term with a comma; for example, if Emma A. Example wrote a book titled *1001 Tips for Best Use of Your Kindle*, she might enter: Emma A. Example, Emma Example, 1001 tips Kindle. She might also want to receive alerts about her topic in general to help her stay on top of the latest news and developments. In that case, she might also include terms such as Kindle, eReaders, eReading devices, and eBooks, for example.

Google allows you to create up to 1,000 alerts. The "Type" drop down menu allows you to select from News, Blogs, Web, Video, Groups, and Comprehensive (which includes all of the above). The Type field determines where Google pulls information for your alert. You can also specify how often you want to receive alerts, how many results should be displayed in each alert (select from 10, 20, or 50), and the e-mail address where you would like to receive your alerts.

Google Analytics

For those brave souls among you who do not fear the mildly technical, you may be interested in Google Analytics. Located at **www.google.com/analytics**, this tool is especially for people who have access to a Web site's coding, functionality, setup, and/or back-end access. If you want to track how effective your marketing campaigns are in terms of Web traffic fluctuations, this tool is good for analyzing Web traffic, conversion rates (how many of the people visiting your site are buying your products or services), the effectiveness of your marketing campaigns, and more. Analytics does not require a great deal of technical knowledge, but does require access to a Web site, as you will have to embed tracking code in the HTML of your site. It is, however, friendly to most users. More sophisticated features are available for tech-savvy users. The Google Analytics family of products includes Website Optimizer and AdWords.

Twitter and Social Media Marketing

Twitter basics

Twitter is a simple yet powerful micro-blogging application whose power stems partly from the fact that, as **www.mashable.com** reported in September 2009, there are 12 million users signed up for the service. Who are those 12 million users? According to **www.istrategylabs.com**, they are mostly between ages 18 and 34, a demographic accounting for 47 percent of users; educated, with 46 percent of users having collegiate degrees and 17 percent with post-graduate educations; and more or less evenly distributed across gender lines, with female users coming in at 53 percent, and males at 47 percent.

Created in 2006, Twitter (**www.twitter.com**) is based on the simple question, "What's happening?" — a question that appears at the top of the page when a user is logged in. Responses can be up to 140 characters in length. The streaming feed shows the "tweets" of everyone you are following. To follow a person or business, search for them by clicking the "Find People" link at the top of the page. You can search for people using four methods: by browsing suggestions, finding friends through your existing e-mail contacts, inviting them to join by e-mail, and looking up their Twitter username. Another great way to find new contacts is by looking at who your current contacts are following. You can see anyone's follow list simply by clicking on his or her username, and then clicking the "following" link just below the number of followers; this brings up the queue of everyone he or she follows.

Twitter is a good way to keep up with what is going on with your contacts, which is also the primary reason it was developed, according to the site's About Us page. It is easy to help spread information posted by those whose tweets you follow, simply by "retweeting" their tweets. If someone posts something informative or entertaining, you can re-tweet the post by hovering your mouse over the tweet, then clicking the "Retweet" option that appears in the lower right of the post cell. This will show up as "RT." The tweet is then visible to all of your followers. Another way of marking tweets you like is to "favorite" them. Do this by clicking the star that appears in the tweet's cell when you hover your mouse over it. Anything you "favorite" will appear in a separate queue, accessible through the "Favorites" link in the right navigation area.

To view any tweets that have mentioned your username, or to view posts where others have replied to you, click the @YourName link, located in the right navigation area to see the feed. You can direct message other "tweeps" or "tweeple," as they are coming to be called, either by navigating to the "Direct Messages" link on the right, or by clicking on the person's username; once you are on their page, click the drop down menu with the gear icon, and choose the option to direct message that person. That menu also includes options to "unfollow," to block, to mention (you can also mention simply by including @UserName in a tweet), and to report for spam.

Another way to stay on top of hot topics is by making use of trending topics. These are located in the right navigation area, under "Trending Topics." Most of these begin with a hash tag (#), and the topics that appear in the queue are the most popular Twitter-wide, at least at the moment. To see what people are talking about under these topics, click a link in the list to see the topic feed. To respond to a topic, simply type your message, include the topic and hash tag in your response, and it will appear in that topic's feed. To search for other topics, such as those related to your field or industry, type a term in the search field and hit enter. In general, try to follow as many like-minded people as you can find, and pay attention to those hash tags. You may discover a new trending topic that relates specifically to your area of interest.

Twitter brings in so much information from so many people at once, that the "powers that be" have devised ways of helping their users organize the information. One of the ways to do this is to create saved queues by marking posts as favorites, and by creating lists. "Lists" has its own area in the right navigation area, and is located directly under the search field. Lists are ways of organizing the people you follow. For example, you might decide to create a list for writers by clicking "New List" and titling it "Writers." Lists are especially useful if you follow a lot of people, or if the people you follow tend to be divisible into separate categories, like "Newspapers," "Musicians," "Friends," and "Economists," to name a few examples. If you are just getting started on Twitter, and are looking for like-minded people to follow, the best place to begin is by looking at the lists of people with similar interests as yours. You can choose to follow the same people, and add them to your list. The easiest way to add new users to your lists is to access your following queue by clicking the "following" link at the top of your page. In the queue, and after creating your lists, add users to them by clicking the drop-down menu with the bullet list icon and checking the appropriate checkbox where you want the person to be listed.

Using Twitter as a marketing and promotional tool

Authors, actors, musicians, deejays, filmmakers, plumbers, students, business executives, and everyone in between is using Twitter. The reason is apparent: If you are marketing a product or service, or are a professional who wants to share information about their services, or if you want to get the word out about anything at all, the simplicity and benefits of Twitter's information sharing capabilities seem almost too good to be true. When you are thinking of promoting your work as an author or marketing your services as a professional, it is especially important to develop a following of like-minded people and professionals in your industry, and of course, by end-users and potential readers. Generating a following — though it takes time — is fairly simple: follow and you will be followed. There are a few caveats to that, however — you should tweet regularly and mindfully.

As with a blog or Web site, you demonstrate your dedication and professionalism by updating regularly; it is the same with any service you use, even Twitter. Find something to say at least once per day, and you (and by translation, your products and services) will stay in the public eye, or at least in the eyes of those who follow you. Tweet interesting, informative, and valuable or even humorous information that people will want to read and share. People have tired of reading marketing copy. They have developed what is known in the industry as advertising "blindness," so your marketing has to be a bit more clever and should make you seem more approachable and human — not just some robot hiding behind an imposing corporate wall.

As an author, your primary goal should be to promote your professionalism and industry knowledge, and by translation, your book. If you are a fiction writer, consider tweeting quotes from authors once per day. Authors of all stripes can tweet writing tips, and informational excerpts from their books that will (hopefully) entice readers to purchase their work. Look at the tweets of other authors on Twitter to see what they are talking about (hint: when looking up celebrity accounts, look for the "Verified Account" symbol, which is a white check mark inside a blue cloud that is provided as evidence of authenticity; see "@MargaretAtwood" for an example of a verified author account). Retweet anything you find useful or interesting, as this can help increase your visibility and encourage positive relations with others in your field. The positive benefits with regards to re-tweeting have a way of building on themselves; much like sales of multiple titles and positive feedback encourage your sales ranking in the Amazon store.

Besides useful information, you can also include links in your tweets, which is a helpful way to drive traffic to your Amazon book pages and to your blog, Web site, and any interviews, articles, or guest blogs you may be able to link to. Bit.ly℠ (**http://bit.ly**) is a useful tool for shortening long URLs, and for that reason it is used often in Tweets. But Bit.ly is also multifunctional, and tracks how many people have clicked on the links you have entered into the main shortener field. The history field on the homepage consists of a list of your links; clicking on the number in blue to the left of the link invokes data on how many people have clicked on the link, their country of origin, and any other mentions of the same link, such as in re-tweets.

Also, consider hosting "events" like tele-seminars, which can be as simple as you video-recording yourself giving helpful advice or a mini how-to lecture in your area of expertise. You can post the video on YouTube, or better yet, on your Web site or blog, and link to it through Twitter. Let people know well in advance when the event will be, and tweet about it at least a week beforehand, and about once or twice per day.

But be careful how you approach what is obviously a promotion; including a link along with a useful tip you will be offering or preview should do just fine. The "event" part of the arrangement can be as simple as you posting the video in its entirety on a certain date and time, making yourself available for chat, and encouraging a designated time during which you will facilitate a Q&A.

Look at Twitter as a tool that is helpful for generating awareness and appreciation of your writing and of yourself as a writer — but remember that it is also important to stay relatable. If you go by the "80/20 rule" that Meredith Liepelt writes about in her Sept. 21, 2009 article for Site Pro News (**www.sitepronews.com**), titled "Simple Twitter Tips that Attract Clients and Partners," you would post two business or professionally related items for every eight tweets you post concerning non-business-related topics. Do tweet personal things about yourself that will be of interest to your target audience, which is whoever you think will be reading your work; also target those you hope will read your work, and it is likely that they will follow.

It should go without saying that you should always use discretion when considering what to post. Know that what you post has a long shelf-life, and that there are search engines designed specifically for social media outlets such as Twitter, and even some that solely search Twitter posts (a comprehensive list of these and other social media tools is provided in the following section). Before posting, also consider that what you say goes a long way toward generating your public image. It may be helpful to keep in mind that hobbies, children, pets, and food tend to generate the most positive response over any other topics, so you may consider tailoring your personal tweets accordingly, if possible.

Social Media Tools

In the age of *Web 2.0*, a term that has come to stand for the increasing attention being given to online interactivity, social media is the current of marketing. It makes sense, if you think about it — position yourself where everyone socializes, and people are bound to notice you. But remember that term, ad blindness? It means that you have to be as careful — across the board — about how you position and market yourself and your image as a writer as you do when using Twitter. Marketing your eBook has never been easier, but that does not mean that the marketing should be done without careful thought. Done properly, social media marketing can be a powerful way to build your reputation as an author and to get your eBook in front of as many eyes as possible.

It is helpful to go in with a plan, or at least the idea of a campaign. Marketing your work as an author is somewhat different from business marketing, but it is so primarily in name — the principles remain the same. A business is trying to sell a product or service, and in a way, so are you. Even if you do not have a corporation or publishing company behind you, learn to think like a savvy marketer and your sales will see an increase. And take it from someone who knows — marketing is not exactly second nature to a writer who enjoys reading Wallace Stevens and T.S. Eliot in her spare time, but it is worth learning if you want your writing to become something that is working for you.

An awareness of how you can use social media to your benefit is important, and there are some tools that can help you along as you work to build an online identity and get the word out about your writing. You may also find it important to monitor the buzz and stay abreast of your online reputation. Some social media tools that you may find useful are:

- Social Oomph: **www.tweetlater.com** — This service allows you to set up Tweets in advance and schedule them to appear at a certain date and time. Free and paid options are available. Free option includes an e-mail newsletter that lists tweets with keywords and key phrases you specify.

- Search engines for tweets: **http://topsy.com**, **http://backtweets.com**, and **www.twazzup.com** — These sites track your retweets and any comments about you; type in your username to search.

- Mentions: **http://socialmention.com/alerts** and **http://sitemention.com** — Similar to Google alerts, but tracks social media according to terms you define.

- Conversation indexes: **www.backtype.com** — Search engine that indexes conversations from social media outlets, such as blogs and other social networks.

- Specialized results: **http://addictomatic.com** — Provides you with a specialized results page when you type in search terms. Searches news sites, blogs, Google, YouTube, Flickr™, Digg™, Technorati, Topix™, Ask, and others.

- Buzzoo: **http://buzzoo.net/ui** — Tracks several Web sites for hot topics.

- Blog pulse: **http://blogpulse.com** — Search engine for blog data.

- Tracker: **www.boardtracker.com** — Search engine for forums.

- Monitoring tools: **http://sm2.techrigy.com/main** and **www.trackur.com**— Social media monitoring tools. Free and paid versions.

- Reputation tracker: **www.filtrbox.com**, **www.brand-seye.com**, **www.reputationdefender.com**, and **www.brandwatch.com** — Tracks online reputation through Twitter, online news sources, and others. Free and paid versions.

- Alert organizer: **www.alertrank.com** — Organizes Google alerts in a daily report spreadsheet.

Amazon Community Forums

Do not rule out the Digital Text Platform Community Support forums, located at **http://forums.digitaltextplatform.com/dtpforums/index.jspa**. While the communication from Amazon was once minimal, it is beginning to gain some ground as the Kindle platforms develop. New posts are now made in the forums by administrators when a new feature or development is added or when something in the policies or procedures changes. The forums are also useful for networking with other writers and publishers who may have already experienced an issue you are facing; performing a search on your topic before posting a new thread may reveal similar issues that have already been resolved. Popular Kindle authors and bloggers like Stephen Windwalker frequent the forums and often step in to provide useful information for users who are struggling with certain topics.

CASE STUDY: TECH EXPERT PREDICTS CHANGES IN KINDLE

Paul Biba
Coeditor of Teleread,
www.teleread.org
www.palmaddict.typepad.com
www.gpspassion.com

CLASSIFIED CASE STUDIES
™
directly from the experts

As the coeditor of TeleRead and an international corporate lawyer by trade, Paul Biba is well known for his TeleRead contributions and as iPhone editor for PalmAddicts and chief U.S. equipment reviewer for GPS Passion. Paul lives in New Jersey and has been interviewed for articles about eBooks in the New York Times, The Guardian, and the Christian Science Monitor.

New York Times eBook article: **www.nytimes.com/2007/08/09/ technology/circuits/09basics.html?_r=2&fta=y**

The Guardian article: **www.guardian.co.uk/technology/2009/ dec/23/amazon-kindle-ebook-sales-guessing**

Christian Science Monitor article: **www.csmonitor.com/The-Culture/Arts/2009/0130/p13s01-algn.html**

As Co-Editor of the TeleRead blog, which is the oldest English language publication dealing with eBooks, eReading, and ePublishing, I am responsible for creating most of the posts; soliciting, editing, and publishing posts and articles by others; all administrative matters relating to the WordPress installation; moderating comments; and deleting spam.

The time for eReaders seems to have come. Amazon is reporting that it sold more eBooks than paper books over the 2009 Christmas season. People see a need for something small and light that can carry around an entire assortment of books and periodicals. This is a need that has been, at least partly, fulfilled by the availability of the readers themselves. I do not think most people saw eReaders as a need, but now that they are becoming mainstream, people are seeing that the eReader is something that they always wanted but did not know it. EReaders are also timely in the sense that huge numbers of people of all ages have now become accustomed to computers and smart phones, so the very idea of an eReader is something that they do not find puzzling or intimidating. The "computer education" of the public was a prerequisite for this technology to take off.

The technology is in its infancy and is certainly not the best at the present time. Future developments will focus on battery life and screen technology. Improvements in these areas will drive the types of eBooks that will be read in the future. Presently art books, for example, are not suitable for eReading, as color screens are not available. We need longer battery life, and color and displays with a faster refresh rate. These are not easy things to develop, so it will be at least two years before we reach any sort of optimum hardware configuration.

EPublishing is driving the industry nuts, to be plain and simple. Publishers are conservative and are not used to dealing with change. As a result they have no idea how to respond to the pricing, licensing, piracy, and other issues that eBooks are creating. As a result, they are flailing around trying one solution after another, but have no game plan or overall strategy for where they want to go. Their responses have been ad hoc and driven by outside forces rather than being self-generated and part of an overall strategy. To further complicate matters, the rise of eBooks is cutting out the distributor and reseller network that publishers are used to dealing with. As a result, they are now starting to deal directly with consumers ☐ something they never had to do before, and they do not know how to do it. EPublishing will probably force the most major changes the publishing industry has seen in 100 years.

During the course of writing this book, so many changes happened regarding the Kindle and ePublishing and eReading devices that I found myself continually going back to make alterations and write updates to the information. By the time this book gets into your hands, there are bound to have been still more changes, however large or small, important or minor. But of course, that is the nature of technology, which is responsible for moving the previously stable publishing industry forward as fast as new devices and systems can be dreamed up. And that is tremendously exciting.

On Jan. 27, 2010, Kindle's chief competitor, the long-awaited Apple iPad, was launched. It was the moment many Mac users, tech-geeks, and ePublishing enthusiasts had been waiting for. Some automatically assumed that the iPad was a "Kindle killer," and others took more of an analytical, wait-and-see type of approach. While at the time of this writing, it is still a bit too early to tell what effects, if any, the Apple ePublishing application iBooks will have on the sales of Kindles and books from the Kindle Store, all the signs say that the Kindle is here to stay.

Conclusion

For one thing, the iPad does not use e-Ink, but instead uses a backlit screen. Getting rid of the need for backlighting, which can make it difficult to read in direct sunlight or bright lights, is the entire reason for e-Ink. The iPad functions a lot like a mini notebook computer, with its Web browsing capabilities, sophisticated game and graphics interfaces, and basic PDA-like functionalities, such as the calendar, maps, and photo organizers. It is priced considerably higher than the Kindle, at a debut price of $499 (the latest generation Kindle costs $259). For $10 less, at $489, you can purchase a Kindle DX with global wireless capabilities and a non-backlit, e-Ink screen.

Will people buy the iPad and read eBooks on it? Yes, and many of them will go in wanting the latest device made by Apple or because they are gadget enthusiasts, not because they feel the iPad is a superior eReading device. According to a Jan. 28, 2010 article on *Times Live,* senior Motley Fool technology analyst Seth Jayson said, "I have my doubts that people at Amazon are too worried about this." User comments on several similar articles echoed this idea, citing the iPad's focus toward games, video, Web, and graphics over innovation for eBooks. Also telling was the fact that the unveiling of the eBook application on the iPad came after the introduction of the other features. And if any of this rings bells of the by-now decades-old Mac versus PC debate, well, then perhaps you are picking up on all the key signals.

Just prior to the launch of the iPad, Kindle added global wireless capabilities to the DX, allowed international publishers to sign up for a DTP publishing account by removing the U.S. bank account requirement, allowed publishers to select through the DTP whether to enable DRM (digital rights management), announced plans to increase the royalty rate to 70 percent beginning June 30, 2010, and made a handful of other updates to the Digital Publication Distribution Agreement (check the most recent guidelines by typing "digital publication distribution agreement" into the search field at **http:// forums.digitaltextplatform.com/dtpforums/index.jspa**).

And significantly, barely one week prior to the launch of the iPad, Kindle opened up to developers. This was both a wise move and a timely one. The Kindle Development Kit has opened the doors for apps submitted to the Kindle. With iPhone apps in place that allow users of the device to play games and read eBooks on their phones (among many other things), it is clear that the people behind the Kindle obviously understand their competition. The limited beta phase for Kindle developers began February 2010, and according to the Kindle Development Kit for Active Content (just type this into a search engine to pull up the page, which has a lengthy URL), developers who are participating in the limited beta phase will have access to development kit downloads, will be able to test their content, access support for developers, and submit their content. Some who apply will be wait-listed and invited to participate in the program as openings become available. Interestingly, developers will receive the larger half of a 70/30 percent revenue split, net of a $0.15 delivery fee. According to a Jan. 21, 2010 article on Inside Tech, Kindle anticipates apps like cookbooks, word games, and weather and travel guides.

Aside from all the innovative features in the works for the Kindle — and even apart from the latest upgrades and developments to the technology — some of the most significant impacts the device will continue to have are on the publishing industry. On one hand, there is Apple, who at the launch of its iPad, announced partnerships with some of the biggest publishers in the industry — namely, Penguin, HarperCollins, Hachette Book Group, and Simon & Schuster MacMillan. And on the other hand, there is Amazon, whose platforms allow authors and publishers of books, blogs, newspapers, and magazines equal access to the same store, where all are on level footing, and all are given equal opportunity to succeed and profit. Now, this is starting to sound less like the Mac-PC debate, and more like the traditional publishing industry, now ruled by media mega-conglomerates (CBS Corporation owns Simon & Schuster Macmillan and News Corp owns Harper-Collins), versus independent publishing. But Kindle has a leg up over the iPad for many authors, as its DTP is open to all users, while the iPad currently remains closed to all but the large publishing companies with whom it has struck deals.

EBooks and their methods of distribution are upsetting the financially driven ways of the traditional publishing industry. In a market where publishing companies and their gatekeepers are relied upon less and less, bottom lines (the ones beginning with dollar signs) are beginning to crumble. Many people are watching the publishing industry with keen interest, and looking to the Kindle as the device that started a fire.

Web sites and/or blogs related to eBooks, digital publishing, and other author resources

- PEN:
  www.pen.org

- Society of Children's Book Writers & Illustrators:
  www.scbwi.org

- The Authors Guild:
  www.authorsguild.org

- The Graphic Artists Guild:
  www.graphicartistsguild.org

- Google Books:
  http://books.google.com

- Mobipocket eBookBase:
  www.mobipocket.com/ebookbase/en/homepage/default.asp

- Publetariat: People Who Publish:
  www.publetariat.com

- Publishers Weekly:
  www.publishersweekly.com

- The Creative Penn:
  www.thecreativepenn.com

- Author 2.0:
  http://author2zero.com

- John August Screenwriting Tips:
  http://johnaugust.com

- The Reading Edge Podcast:
  http://thereadingedge.com

- TeleRead: **www.teleread.org**

Web sites and/or blogs related to Kindle

- Amazon DTP:
  http://dtp.amazon.com

- Kindle Publishing for Blogs:
  http://kindlepublishing.amazon.com

- The Kindle Chronicles:
  www.thekindlechronicles.com

- A Kindle World Blog:
  http://kindleworld.blogspot.com

- Kindle Nation Daily:
  http://kindlehomepage.blogspot.com

- Kindle Boards Blog:
  www.kindleboards.com

Common HTML terms:

HTML: Stands for HyperText Markup Language.

Class attribute: An attribute whose value is common to an entire class of objects and not just to a value particular to each instance of the class.

Attribute: A named property of a class that describes a data value held by each object in the class.

Tag: Indicator used to set sections apart, such as bold text, italics, underlined text, and headings found within the <>. For example, a paragraph tag is "p," represented as <p>.

Block-level element: A page element that defaults as a block on the page, and will begin new line breaks around itself. Defaults to the width of the screen unless you add specifications for a certain width.

In-line element: Unlike the block-level element, the in-line element does not create new lines around itself. Bold

Appendix B

<strong> and italics <em> fall into this category. In-line elements are contained within block-level elements.

CSS: Cascading Style Sheets. Styles are pre-designated page elements that have a certain appearance, depending on how you designate the settings. CSS elements are organized in hierarchies.

HTML Basics

- Tag order is important. When more than one tag is written, it should appear like this:

  > <strong><em>Strong and emphasis equal bold and italic</em></strong>

- Make sure that you "end" your tags by including the / (see the example above). This is usually referred to as the open and close tag format.

- Two tags are stand-alone, and do not require the forward slash closure. They are:

  > `<HR>` Horizontal Rule: gives you a horizontal line across the page. This is a self-closing tag, and can be written as `<hr />`.

  > `<BR>` Break: a line break moves the text to the next line. Use in place of hitting "Enter."

- At the beginning of every page, designate the 'doctype' such as:

```
<!DOCTYPE    HTML    PUBLIC    "-//W3C//DTD
HTML    4.01//EN" "http://www.w3.org/TR/
html4/strict.dtd">
```

This tells the browser which version of HTML you are using. If you use a Web design program, such as Dreamweaver, this is automatically inserted into every document for you. If you use CSS (cascading style sheets), this header is particularly important. The exclamation mark at the beginning of the tag indicates a comment; comments are embedded within the page and do not display to the viewer.

Begin every page with <HTML> and end with </HTML>

The second tag on every page will always be <TITLE>, closed with </TITLE>

Paragraph <p> and line break <br> tags are used to create new paragraphs and move the text to a new line, respectively, and do not need closing tags.

The following is a simple sample page:

```
<!--DOCTYPE HTML PUBLIC "-//W3C//DTD HTML 4.01//
EN" "http://www.w3.org/TR/html4/strict.dtd"-->
<html>
<head>
    <title>A simple HTML document.</title>
</head>
<body>
    <p>Part <strong>One</strong><p>
```

```
<br>

<p>The HTML Tutorial</p>
```
```
</body>
```
```
</html>
```
```
<!-- This is the end of the document -->
```

- Use a plain text editor to write HTML, like Notepad, Simple Text, or WordPad. When you save the document, the default file extension is .txt. A plain text file saves the text only and does not convert or change it like a word processor would.

- If you use a word processor to write your HTML, the most important thing to remember is that every time you save your document, you must go to File and choose Save As and save the document as Text or ASCII Text DOS on a PC or Text on a Mac. If you do not choose Save As > Text, you will save your file as the default word processing document, which will interfere with your HTML.

- When you save your document, name it example.htm or example.html. Generally, use .htm if you are running on a system that uses Windows 3.x and .html on a Mac or for Windows 95/98 or later. This is important to tell your computer what type of file this is.

- Check out the competition. It is a good idea to look around at other documents, especially when you are first learning HTML. Look around for something you like, and then check behind the scenes to see how it was done. To view a Web page's source code, go to View > Source.

- CSS takes care of elements that used to be handled by HTML, such as text size, which used to be designated with the <FONT SIZE> tag and is now handled with the style attribute, like so:

```
<p style="font size: 150%;">This repre-
sents an in-line CSS style.</p>
```

- CSS is written: Property: value; a property could be anything like the color, border, or font size, and the value is what you want the property to assume. These are separated with a colon and ended with a semicolon to signify the completion of the property.

Common HTML Tags:

TAG	NAME	EXAMPLE
<!--	Comment	<!-- Comments do not appear in the visible part of the document, only in the HTML -->
<a>	Anchor	<a href="http://www.amazon.com/">Visit Amazon</a>
<b>	Bold	<b>Text here will show as bold.</b>
<strong>	Bold	<strong>Text here will show as bold.</strong>
<i>	Italics	<i>Text here will show in italics.</i>
<em>	Italics	<em>Text here will show in italics.</em>
<big>	Big text	<big>Increases the size of the text. To increase the size of the text further, use two of this tag side by side.</big>

TAG	NAME	EXAMPLE
` `	Line break	`This is the first line.` ` ` `This is the second line.`
`<body>`	Body of document	`<body>The entire contents of your Web page go between these tags.</body>`
`<center>`	Center	`<center>Everything between these tags will be centered.</center>`

`<dd>`	Definition description	`<dl>`
		`<dt>Definition Term</dt>`
		`  <dd>Definition of the term.</dd>`
		`<dt>Definition Term</dt>`
		`  <dd>Definition of the term.</dd>`
		`</dl>`

`<dt>`	Definition term	See example above.
`<dl>`	Definition list	See `<dd>` example above.
`<font>`	Font	`<font face="Calibri" size="3" color="#003300">Sample text</font>`
`<head>`	Document heading	`<head>Elements here describe your HTML document, and are visible only in the HTML.</head>`
`<H1>`	Heading 1	[Headings are commonly used for Web pages; browsers recognize headings as important page elements.] `<H1>Heading 1 is the largest, about 24-point font.</H1>`
`<H2>`	Heading 2	`<H2>Heading 2 is the second largest, about 18-point font.</H2>`
`<H3>`	Heading 3	`<H3>Heading 3 is the third largest, about 13.5-point font.</H3>`
`<H4>`	Heading 4	`<H4>Heading 4 is the fourth largest, about 12-point font.</H4>`
`<H5>`	Heading 5	`<H5>Heading 5 is the fifth largest, about 10-point font.</H5>`
`<H6>`	Heading 6	`<H6>Heading 6 is the sixth largest, about 7.5 point font.</H6>`

`<hr>`	Horizontal rule	`<p>Web page contents</p>`
		`<hr />`
		`<p>Web page contents</p>`

`<img>`	Image	`<img src="FileOnYourComputer.gif" width="12" height="12" border="1" alt="text that describes image" />`

TAG	NAME	EXAMPLE

`<li>`	List item	[Bullets can appear as a filled-in circle, open circle, or square:]
		`<menu>`
		`  <li type="disc">List Item 1</li>`
		`  <li type="circle">List Item 2`
		`  </li>`
		`  <li type="square">List Item 3`
		`  </li>`
		`</menu>`

`<link>`	Link	`<head>`
		`<link rel="stylesheet" type="text/`
		`css" href="style.css" />`

`<menu>`	Menu	Menu tag precedes a list. See `<li>` example above.

`<meta>`	Meta	`<meta name="Description"`
		`content="Your site description`
		`here">`
		`<meta name="keywords"`
		`content="Keywords describing site`
		`here">`

`<ol>`	Ordered list	[Instead of bullets, list items can be listed by lower case roman numerals:]
		`<ol type="i">`
		`  <li>List Item 1</li>`
		`  <li>List Item 2</li>`
		`  <li>List Item 3</li>`
		`  <li>List Item 4</li>`
		`</ol>`

`<p>`	Paragraph	`This is an example`
		`sentence.<p>This sentence is on`
		`the second line.`
		`<p align="left">Use this tag to`
		`left-align the text that follows.`
		`<p align="right">Use this tag to`
		`right-align the text that follows.`
		`<p align="center">Use this tag to`
		`center the text that follows.`

`<small>`	Small text	`<small>Creates small text. Use two`
		`of these tags side by side to make`
		`the text even smaller.</small>`

`<strike>`	Strikethrough	`<strike>Strikes through this` `text.</strike>`
`<title>`	Document title	`<title>Title of the HTML page, is` `displayed in browser bar.</title>`
`<u>`	Underline	`<u>This text will be underlined.</` `u>`

TAG	NAME	EXAMPLE
<ul>	Unordered list	[In this example, the list items will appear with closed-circle bullet points in front of the list items:] ` ` `<ul>` `<li>List Item 1</li>` `<li>List Item 2</li>` `</ul>` ` ` [In this example, you can control the appearance of the bullets in front of the list items; list items 3 and 4 are offset by one tabbed space from the first and second list items:] ` ` `<ul type="disc">` `<li>List Item 1</li>` `<li>List Item 2</li>` `<ul type="circle">` `<li>List Item 3</li>` `<li>List Item 4</li>` `</ul>` `</ul>`

Sample Press Release

Children's Book Reveals the Importance of Team Sports

New York, NY, October 5, 2010 — Bestselling children's author Jane Writer explores the other side of youth sports in her latest book from Best Books Ever, *Jiminy Jumping Jacks!* Released in October 2010, *Jiminy Jumping Jacks!* is available for $14.95 from Barnes & Noble Booksellers, Borders Books, Amazon.com, and other fine booksellers nationwide.

Writer's picture book promises to make an impact on children everywhere who struggle with weight problems. These children often suffer from low self-esteem and feelings of inadequacy, especially when it comes to participating in team sports. While the situation may be grim for a portion of American's population, Writer makes a bleak situation fun and addresses a difficult topic effortlessly. The story of turtle and his friends exhibits a supportive

Appendix C

situation that is sure to be a hit with parents, teachers, and children's librarians.

Writer has taught physical education to children for 10 years, and has seen firsthand the positive effects of team sports on overweight children. Her program, "Fitness is Fun," has helped more than 75 children reduce their weight and see measurable health benefits. And Writer certainly has an audience — according to a recent study, one out of every three American children is overweight or obese. Says Writer, "Educating children about eating right and the importance of exercise is vital. But children don't have to know that — I hope my book makes it fun for them."

For additional information, contact Netty Networker at (555) 555-1234.

About Jane Writer: The award-winning author of *This is My Bus, Jimmy's Magic Day*, and *The Dog that Wouldn't Stop Running*, Jane Writer is a dedicated and talented writer for children. She has published stories in *Highlights for Children, American Girl, Muse, Cricket*, and *Nick*. When she isn't writing or reading to kids, she loves to go horseback riding and take picnic lunches with her three children.

CONTACT INFORMATION:

Netty Networker
XYZ PR Firm, Inc.
(555) 555-1234 (voice)
(555) 555-2345 (fax)
netty@xyzpr.com
www.xyzpr.com

Sample Review Query

Jane Writer
1234 Show-Don't-Tell Lane
Boston, MA 02101

January 1, 2011

Publication of the Month
Attn: Ricky Reviewer
1234 Bowery Blvd.
New York City, NY 10010

Dear Mr. Reviewer,

My book was recently released from Best Books Ever. I am writing to ask if you would be interested in reviewing it. *Jiminy Jumping Jacks!* is the tale of a young turtle's struggle to live up to the athletic triumphs of his best friends, Rabbit and Fox.

With an increasing focus on keeping kids healthy, parents, teachers, and librarians will appreciate the values represented in the adventures of the three friends. The book

Appendix D

is fun for kids to read and not only shows the importance of eating right and exercise, but also deals with an experience many kids can relate to. For children whose athletic abilities are not quite at the level as those of their friends, participating in sports can be disheartening. Children who are overweight often struggle with body image and feelings of social inadequacy, and turtle's interactions with his friends are taken from real-life examples.

I have taught physical education for 10 years, and have seen firsthand the effects that feeling inadequate has on children's self-esteem. Not many children's books address this topic, and I hope that mine call fill a void while showing children that they can turn this difficult situation around through both perseverance *and* fun.

Please let me know if you are interested, and I will gladly send a complimentary review copy your way. Thank you for your time.

Sincerely,

Jane Writer

Sample Review Slip

Best Books Ever Presents for Review

Title:	*The Legacy of Thomas Edison: More than Just the Inventor of the Light Bulb*
Author:	Cynthia Reeser
Edition:	First
Number in Print:	100,000
CIP/LCCN:	1234567890
ISBN:	1-23456-789-0
Pages:	64
Cover art:	Illustration by Cynthia Reeser
Price:	$14.95
Season:	Fall 2010
Publication date:	October 2010
Rights:	a. Subsidiary: Book club, paperback
	b. Syndication

Please send two copies of your review to the address below:

Best Books Ever
Public Relations Department
1234 Bindery Blvd.
New York City, NY 10010

Tel: (555) 555-1234; Fax: (555) 555-2345
Info@BestBooksEver.com
www.BestBooksEver.com

Edwards, Jim and Joe Vitale, *How to Write and Publish Your Own eBook in as Little as 7 Days*, Morgan James Publishing, Garden City, New York, 2007.

Greco, Albert N., *The Book Publishing Industry*, 2nd ed., Routledge, New York, 2009.

Hicks, Michael R., *Publish Your Book on the Amazon Kindle: A Practical Guide*, Imperial Guard Publishing, 2008.

LaRoche, J. M., *Build a Kindle Edition*, HTMPublishing, 2008.

Reeser, Cynthia, *How to Write and Publish a Successful Children's Book: Everything You Need to Know Explained Simply*, Atlantic Publishing, Ocala, FL, 2010.

Weber, Steve, *ePublish*, Weber Books, 2009.

Windwalker, Stephen, *The Complete User's Guide to the Amazing Amazon Kindle*, Harvard Perspectives Press, Arlington, Mass., 2008.

Bibliography

Windwalker, Stephen, *The Complete Step-by-Step Guide to Publishing Books, Articles & Other Content for the Amazon Kindle*, Harvard Perspectives Press, Arlington, Mass., 2008.

Links to Online Kindle Resources Used

Adams, Tim, "Will e-books spell the end of great writing?," *The Observer*, 6 December 2009. **www.guardian.co.uk/ books/2009/dec/06/books-ebooks-technology-computers-society**

American Library Association, "Google Book Settlement: 2 page Super Simple Summary." **http://wo.ala.org/gbs/wp-content/ uploads/2009/01/Google-Book-Settlement-2-page-Super-Simple-Summary.pdf**

Burns, Joe, "HTML Primer," *HTML Goodies*. **www.htmlgoodies.com/primers/html/article.php/3478131**

Butler, Brandon, "The Google Books Settlement: Who Is Filing and What Are They Saying?," *American Library Association*. **http://wo.ala.org/gbs/wp-content/uploads/2009/10/ The-Google-Books-Settlement-Who-Is-Filing-And-What-Are-They-Saying.pdf**

Cane, Mike, "Apple Will Break Open the Digital Floodgates," *The eBook Test*: Blog, 2 Nov. 2009. **http://ebooktest.wordpress. com/2009/11/02/apple-will-break-open-the-digital-book-floodgates**

Internal Revenue Service, "Business or Hobby? Answer Has Implications for Deductions," 18 April 2007. **www.irs.gov/ newsroom/article/0,,id=169490,00.html**

June, Laura, "Apple reveals iBook Store and app for the iPad," *engadget*, 27 Jan. 2010. **www.engadget.com/2010/01/27/ apple-reveals-ibooks-store-and-app-for-the-ipad**

Kunhardt, Jessie, "eBook Sales Shyrocket, Book Sales Up In General," *The Huffington Post*, 17 Dec. 2009. **www. huffingtonpost.com/2009/12/17/ebook-sales-skyrocket-boo_n_396002.html**

Lyons, Jonathan, "Term of the Week: Net Royalties," *Lyons Literary LLC: Tips and Quips on Publishing from a Literary Agent*, 24 Aug. 2007. **http://lyonsliterary.blogspot.com/2007/08/ words-of-week-net-royalties.html**

Manjoo, Farhad, "Fear the Kindle," *Slate*, 26 Feb. 2009. **www. slate.com/id/2212320/**

McCarthy, Caroline, "Amazon's big-screen Kindle DX makes its debut," *CNET News*, 6 May 2009. **http://news.cnet. com/8301-17938_105-10234355-1.html**

McCracken, Harry, "The Technologizer Review: Amazon Kindle DX," *Technologizer*, 12 June 2009. **http://technolo-gizer.com/2009/06/12/kindle-dx-review**

McGlaun, Shane, "New Dev Kit Means Active Content Coming to Kindle in 2010," *Inside Tech*, 21 Jan. 2010. **http:// insidetech.monster.com/news/articles/7144-new-dev-kit-**

means-active-content-coming-to-kindle-in-2010?utm_
source=nlet&utm_content=it_c6_20100126_HIRE

Milliot, Jim, "Kindle Market Share on the Rise," *Publishers Weekly*, 31 August, 2009. **www.publishersweekly.com/article/CA6686612.html?q=kindle+market+share+on+the+rise**

Moritz, Scott, "Apple's Tablet geared to be a Kindle killer," *MSN Money*, 9 Dec. 2009. **http://articles.moneycentral.msn.com/Investing/top-stocks/print.aspx?post=1419197**

Open Book Alliance, Google Book Settlement Updates. **www.openbookalliance.org**

Platt, Judith, "Google Library Project Raises Serious Questions for Publishers and Authors," *The Association of American Publishers*, 12 Aug. 2005. **http://publishers.org/main/Press-Center/Archicves/2005_Aug/Aug_02.htm**

Reed, Harper, "My thoughts on the kindle: 1 vs. 2 and the iPhone app," *Harper Reed: Tech, Mobile...*, 8 March 2009. **www.nata2.org/2009/03/08/my-thoughts-on-the-kindle**

Semuels, Alana, "Amazon's Kindle has a big job: saving the newspaper industry," *Los Angeles Times*, 7 May 2009. **http://articles.latimes.com/2009/may/07/business/fi-kindle7**

Stone, Brad, "Amazon Cracks Open the Kindle," *The New York Times*, 20 Jan. 2010. **http://bits.blogs.nytimes.com/2010/01/20/amazon-cracks-open-the-kindle/?src=twt&twt=nytimesbits**

Strott, Elizabeth, "Amazon says Kindle most gifted item ever," *MSN Money*, 28 Dec. 2009. **http://articles.moneycentral.msn.com/Investing/Dispatch/market-dispatches.aspx?post=1513820&_blg=1,1513820**

Sullivan, Danny, "Making an RSS Feed," *Search Engine Watch*, 2 April 2003. **http://searchenginewatch.com/2175271**

The Authors Guild, "Amended Settlement Filed in Authors Guild v. Google," 13 Nov. 2009. **www.authorsguild.org/advocacy/articles/amended-settlement-filed-in-authors-guild.html**

Times Live, "Publishers embrace iPad, but revolution unlikely," 28 Jan. 2010. **www.timeslive.co.za/scitech/article280873.ece**

W3Schools, "HTML Tutorial." **www.w3schools.com/html/default.asp**

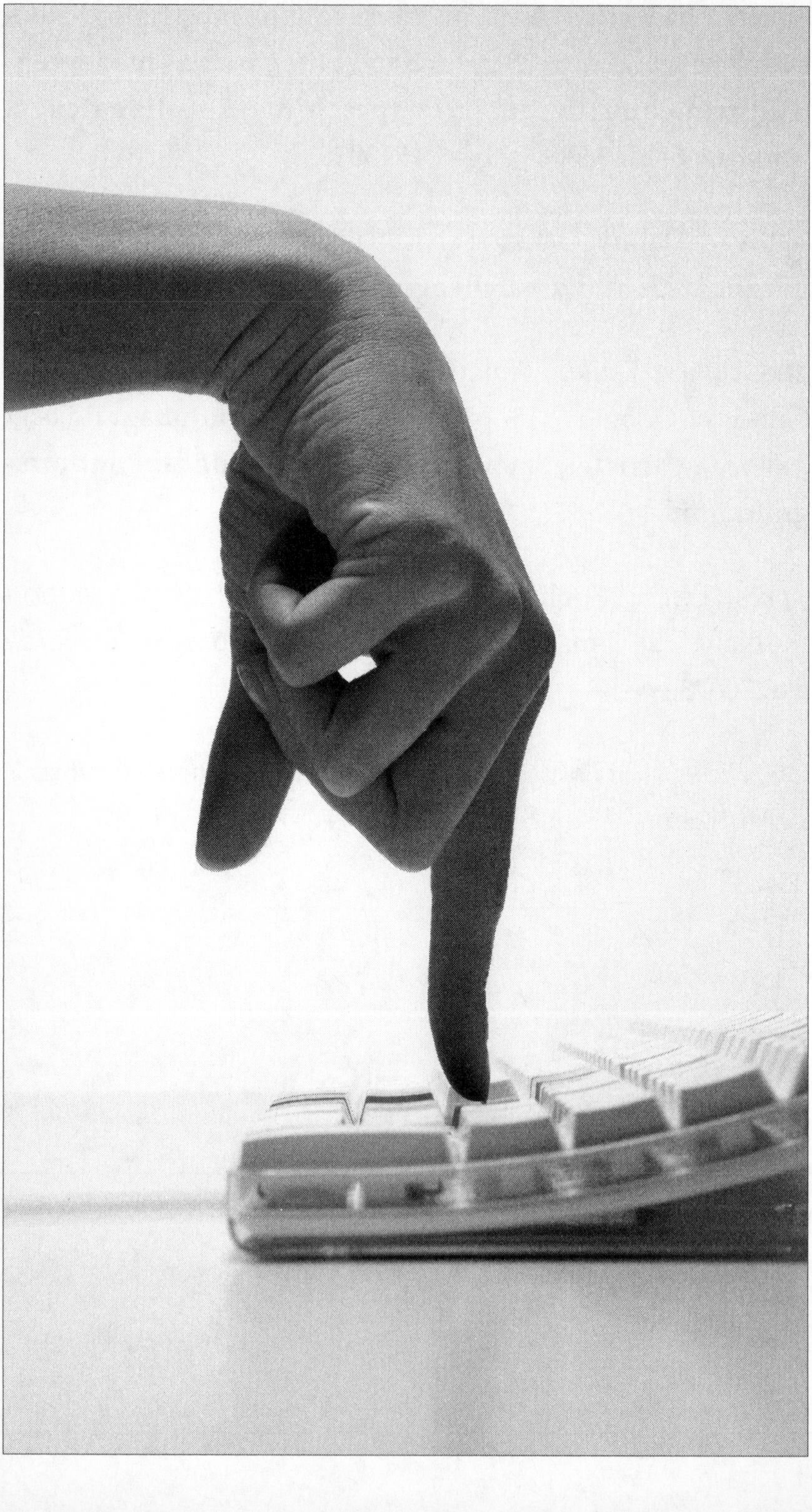

Cynthia Reeser is the editor-in-chief and founder of a quarterly literary journal, *Prick of the Spindle*. She is also a practicing visual artist whose work has been featured on several book covers. Her works of criticism, nonfiction, and poetry are widely published in both print and online media. She is the author of *Light and Trials of Light* (poetry) and *How to Write and Publish a Successful Children's Book* (nonfiction).

Author Biography

Index

C

D

E

F

L

M

N

P

R

S

T

U

W